T0143480

Real-Time and Distributed Real-Time Systems

Theory and Applications

Real-Time and Distributed Real-Time Systems

Theory and Applications

Amitava Gupta
Anil Kumar Chandra
Peter Luksch

CRC Press is an imprint of the
Taylor & Francis Group, an **informa** business

CRC Press
Taylor & Francis Group
6000 Broken Sound Parkway NW, Suite 300
Boca Raton, FL 33487-2742

Printed on acid-free paper
Version Date: 20160108

International Standard Book Number-13: 978-1-4665-9847-8 (Hardback)

Library of Congress Cataloging-in-Publication Data

Names: Gupta, Amitava (Computer scientist), author. | Chandra, Anil Kumar, author. | Luksch, Peter, author.
Title: Real-time and distributed real-time systems : theory and applications / Amitava Gupta, Anil Kumar Chandra, and Peter Luksch.
Description: Boca Raton : Taylor & Francis, CRC Press, 2016. | Includes bibliographical references and index.
Identifiers: LCCN 2015043872 | ISBN 9781466598478 (alk. paper)
Subjects: LCSH: Electronic data processing--Distributed processing. | Real-time data processing.
Classification: LCC QA76.54 .G875 2016 | DDC 004/.33--dc23
LC record available at http://lccn.loc.gov/2015043872

Visit the Taylor & Francis Web site at
http://www.taylorandfrancis.com

and the CRC Press Web site at
http://www.crcpress.com

In memory of Prof. D. Popovic, former professor of the University of Bremen, who has been an architect of an effective Indo–German collaboration in the realm of technology, a teacher for many Indian students in Germany, and a true friend of India.

Contents

Preface

The advent of digital computers during the late twentieth century has not only revolutionized the manner in which computations are carried out, but also the way in which computers can be used to control systems in real life and has given birth to a whole new paradigm termed as *real-time* (RT) systems. The inputs to such a system are usually from the real world and the system processes these inputs to generate outputs that affect the real world within a *finite* time irrespective of the computational load on the processing computer. Further, the massive advancements in the domain of communications have now made it possible to have RT systems performing an action in a coordinated manner over a communication interface, which has modified the RT paradigm further with the evolution of distributed RT systems or DRTS. As an example, one could consider a spacecraft headed toward the moon, the trajectory control for which involves the firing of thruster rockets with specific time deadlines. Missing a deadline in such a case might lead to the spacecraft being lost forever in deep space. Clearly, the requirements needed for the activation of thruster rockets on a spacecraft following a command from the mission control on Earth a hundred thousand kilometers away to meeting a deadline poses a many challenges on the computing resources and the communication channels to ensure timeliness. Necessity being the mother of invention, the requirements of the DRTS applications have paved the way for a whole gamut of techniques to represent, analyze, implement, and finally verify and validate such systems. This book introduces some of these aspects in six chapters that are found between its two covers.

The basis of a DRTS is an RT system. Thus, Chapter 1 begins with the basic concepts related to RT systems, namely, a *task* and its attributes, and the techniques to analyze the schedulability of a set of tasks for achieving a certain RT functionality; it then introduces the principles behind their synchronized functioning. The distributed processing attribute is then added to the RT functionality to cover a scenario where a set of RT systems delivers coordinated functionality through the scheduling of tasks and messages.

Chapter 2 extends the discussion of RT systems presented in Chapter 1 to DRTS by introducing the basic concepts starting with the topology of interconnection, architecture, and properties of a DRTS, and progressively introduces the concepts of time and time synchronization.

The second element in the constitution of a DRTS is an interconnecting network and the communication protocols that can be used to drive it. DRTS applications have led to the evolution of a large number of real-time communication protocols, and a few common protocols are presented and analyzed in detail in Chapter 3. Chapter 3 also presents advancements over the standard switched Ethernet like the multi-streamed TCP/IP, which delivers the reliable communication required, for example, for safety critical DRTS applications. The chapter ends with protocol analysis using a protocol analyzer.

Designing DRTS applications using standard techniques is the theme of Chapter 4. The methodology presented in this chapter is based on a finite state machine (FSM) representation of an RT system. The basic concepts of automata theory are introduced first and then extended to model RT systems using standard FSM representation like the Mealy and Moore machines with practical systems like the coke vending machine. The basic concepts are extended to timed and hybrid automata and finally MATLAB® Stateflow is introduced as a tool for easily reproducing the FSMs as SIMULINK® models.

Chapter 5 illustrates how MATLAB can be used to develop real-time applications and integrate them over a communication network to develop a DRTS. The use of MATLAB allows development of DRTS applications in an easily reproducible manner. The methodology is illustrated using a representative application involving communication between the Earth, a lunar orbiter, and a lander to develop a DRTS. This simulates communication in a deep-space environment with delays taking into consideration the distance and radio-visibility of the communicating components. The chapter ends with an introduction to TrueTime, a MATLAB add-on, which can be used for simulation of protocols and hence DRTS applications in a MATLAB environment.

Finally, Chapter 6 presents the classification of DRTS applications in terms of the criticality and severity of their failures, the metrics in terms of safety and integrity levels, the different stages in their development life cycle, and their verification and validation techniques.

The individual chapters are supplemented by numerical and analytical problems or simulation exercises, which allow better understanding of the concepts. This supports a course format in which lectures are supplemented with tutorials and practical assignments. Such a format is quite common in German and Indian universities. A semester long course may also include a mini project using one of the simulation exercises given in Chapter 5.

Authors

Amitava Gupta is a professor in the Department of Power Engineering at Jadavpur University, Kolkata, India. A graduate in electrical engineering from Jadavpur University, Gupta earned his MTech from the Indian Institute of Technology, Kanpur and his PhD from Jadavpur University. He has served as an engineer in the Control and Instrumentation Division of the Nuclear Power Corporation of India Ltd. and as a member of the high-performance computing group of the Centre for Development of Advanced Computing, Bangalore before switching to academics. He also served as a *Gastwissenscaftler* (visiting scientist) at Technische Universität München, Germany, and as a DAAD *Gastdozent* (visiting professor) at the University of Rostock, Germany and Technische Universität München. His research interests include distributed simulation, distributed real-time systems, networked control systems, and control of nuclear reactors.

Anil Kumar Chandra graduated with a degree in electrical engineering from the Indian Institute of Technology, Delhi. He joined the Bhabha Atomic Research Center, Mumbai, and finally retired as a distinguished scientist and executive director (research and development) of the Nuclear Power Corporation of India Ltd., Mumbai. He worked on the modernization of control rooms and associated instrumentation for forthcoming Indian nuclear power plants. Chandra structured the required systems as distributed architectures with considerable standardization of hardware, software, and human–computer interaction. He also worked on obsolescence management of instrumentation in the older power plants. He contributed to devising the safety guide for review of digital instrumentation and control. He is a senior life member of the Computer Society of India and life member of the Indian Nuclear Society.

Peter Luksch is the chair of Distributed High Performance Computing at the Institute of Computing at the University of Rostock, Germany. Professor Luksch obtained his diploma in computer science from the Technische Universität München, Germany, followed by his doctoral degree and his habilitation. He served as the head of the Parallel and Distributed Applications Research Group of the Institut fuer Informaik of the Technische Universität München and as a visiting scholar at Emory University (Atlanta, Georgia) before joining the University of Rostock. Luksch's research primarily encompasses parallel and distributed computing.

1

Introduction to Real-Time Systems

Real-time (RT) systems belong to a class of computer systems that receive inputs from the external world, process them, and generate outputs so as to influence the external world, within a finite time. The inputs to such a system are usually *events* that occur in the real world, and the time interval between the instant at which such an event occurs and the instant at which the corresponding outputs are generated is termed the *response time* for the system. Thus, for an RT system, the response time is deterministic having a definite upper bound.

The deterministic response time of RT systems makes them suitable for automation-related applications. For example, if one considers a scenario where a thruster rocket is used to control the trajectory of a spacecraft used for planetary explorations, the necessity of deterministic response time becomes clear. The event in this case could be a command to fire the rocket based on the measurement data related to the trajectory of the spacecraft at any time instant (t), and the response time in this case would be the time between the instant at which the command is issued and the instant at which the rocket is fired. Understandably, this interval must have a definite upper bound or else the trajectory of the spacecraft would be completely different. It must, however, be borne in mind that the RT response of a system is associated with the deterministic nature of the response time and not with the magnitude of the response time. It is quite possible that a non-RT system produces a response within a much shorter time span compared to an RT system. For example, if one considers a MATLAB®/SIMULINK® [1] model that simulates the output of a dynamic system due to an input for T s, say, in the non-RT mode, the actual simulation run time may be T', $T' \ll T$ s, while the simulation time in the RT mode should be exactly T s in the ideal case and lie within a deterministic interval $[T - \Delta T, T + \Delta T]$ in the worst case.

As stated earlier, the inputs to an RT system are events that may be asynchronous with respect to the system, that is, the system may not be able to predict the instants at which these occur. In the worst case, the events may be completely unrelated without any *temporal cohesion* or mutual exclusion. Two events are said to possess a temporal cohesion if there exists a defined precedence law for their occurrence. Further, their processing may be completely different with the processing modules having no *functional cohesion* between them. Two processing modules are said to be functionally cohesive if they perform the same or a related set of functions. If one associates an entity called a *task*, which is a sequence of functionally cohesive modules arising out of an event, then it is clear that such tasks, in the extreme case, can be completely independent of each other, without any precedence relationship between them. Thus, an RT system is inherently *multitasking* and RT systems form a special class of multitasking systems.

The rest of the chapter is organized as follows. Section 1.1 introduces the different types of RT systems. Sections 1.2 and 1.3 explain the key concepts associated with RT systems. Finally, Section 1.4 presents numerical and analytical problems.

1.1 Types of Real-Time (RT) Systems

RT systems may be broadly classified into two groups: *hard* RT systems and *soft* RT systems. The terminologies *hard* and *soft* are used to qualify the system based on the consequences of a nondeterministic response. A hard RT system is one in which the output loses its significance if it is not produced within a given response time. A typical example is a fly-by-wire control system controlling projectiles and aircrafts. On the other hand, for a soft RT system, a failure to meet response time requirements results in a degraded response and the *softness* of the system is a measure of its tardiness. A typical example is an RT streaming video display system. For such systems, the importance of the output generated by the system reduces with increased tardiness. The focus of this chapter is mostly hard RT systems, as soft RT systems can be viewed as a special case of hard RT systems with relaxed constraints on response time, and as such the concepts developed for hard RT systems can be easily extended to cover soft RT systems as well.

1.1.1 Embedded Systems

Embedded systems form a class of RT systems that are characterized by a short response time, compact architecture involving a processor with peripherals integrated in a single board often with small usable memory, enhanced robustness and exception handling capabilities, and in most cases low power consumption. To understand the significance of the term *embedded*, the difference between a normal (nonembedded) system architecture and an embedded one should be examined (see Figure 1.1).

As shown in Figure 1.1, in the strict classical sense of the term, an embedded system has the relevant operating system modules that an embedded application uses, embedded within the address space of the application itself. Hence the name embedded system. This arrangement substantially reduces the latency and the memory requirement. In most cases, an embedded

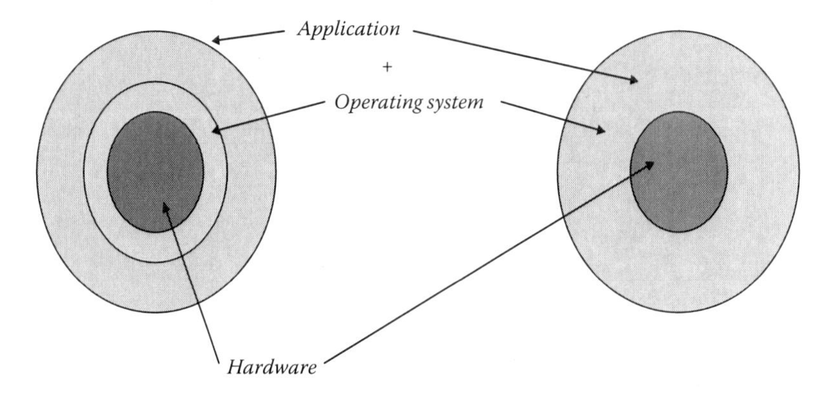

FIGURE 1.1
(**See color insert.**) Difference between an embedded and a nonembedded system.

application is often built around a kernel with basic user interaction facilities, which are often configurable depending on the need of the application. An embedded application usually runs on a specialized hardware called the *target*, which is often configured to run in a stand-alone mode. A target may contain a kernel for running the embedded application (kernel-based embedded system) or may just be able to run an executable, but not the development tools like the compilers and linkers. Such target system development often requires a *development platform* where the executable code is developed for the target, downloaded through a communication link, and stored in a permanent memory device so that the embedded system can be deployed as a stand-alone system or run using the kernel on the target. The development platform, often known as the *integrated development environment* follows an approach, which allows the embedded system developer to pick and choose the operating system modules, for example, drivers that need to be embedded in the application.

1.1.2 Multiprocessor Systems

Multiprocessor architectures are used in RT systems more for providing enhanced reliability than enhanced computing power. Classically, such systems involve multiple *processing elements,* which are used in applications involving RT signal processing, for example, in multimedia applications.

1.1.3 Networked RT Systems

Networked RT systems are a special class of distributed RT systems where the processing elements are connected by a RT communication link. This is covered as a separate topic in Chapter 2.

1.2 Concepts in RT Systems

In this section, the key concepts associated with an RT system are introduced. These concepts define the basic framework for design and analysis of an RT system.

1.2.1 RT Operating System (RTOS)

In the most general case, an RT system consists of a number of tasks programmed to run in an RTOS environment of varying complexity. The general requirements of an RTOS may be enumerated as follows:

- The operating system (OS) must support fixed-priority preemptive scheduling for tasks (both threads and processes, as applicable).
- Interrupts must have a fixed upper bound on latency. By extension, nested interrupt support is required.
- The OS kernel must be pre-emptible.
- OS services must execute at a priority determined by the client of the service.
- A special requirement (for systems with small memory): the kernel should be small.
- User control over OS policies: scheduling and memory allocation.
- Tasks are usually kept fixed in memory and do not use virtual memory.
- Support for multitasking, synchronization, intertask communications, and timer functions.

These requirements ensure that the basic requirement of deterministic response time is met in an RTOS as opposed to a general purpose operating system (GPOS).

An RTOS can be categorized into three basic categories as follows:

1. Kernels for embedded applications—Typical examples are VxWorks [2], MATLAB xPC target, QNX, Windows CE [3], OSE (for mobile phone applications) [4], OSEK (automotive applications) [5], and TinyOS (Wireless Sensor Nodes) [6]. The choice of the kernel depends on the complexity of the application, code size, and response time requirements. Response times as low as 100 μs are achievable with some of these kernels, and typical application code sizes range from 1 KB to 1 MB.
2. GPOS with RT extensions—Typical examples are RT Linux and Windows NT. These RTOSs, like the RT Linux, for example, have an

RT extender, which is implemented as a kernel between the standard GPOS kernel and the hardware. This kernel allows the user to specify RT attributes for tasks and provides functionalities for achieving an RT response. In RT Linux, for example, the RT extender runs standard Linux calls at a lower priority compared to an RT system call. The RT extender filters only RT system calls and no RT system calls are handled by the GPOS kernel. These systems are capable of supporting much larger applications but have larger response times compared to kernels for embedded applications.

3. RT programming languages with runtime systems—A typical examples is RT Java.

This book introduces the basic concepts of RT systems applicable for RTOS categories 1 and 2, since most of the RT applications are built on such RTOSs.

1.2.2 Task

A task may be defined as a sequence of instructions in execution. A task usually interacts with other tasks and uses resources like memory, input and output (I/O) devices, and processors. Two tasks, T_i and T_j, may be related by a *precedence* or a *dependency* relationship. Task T_i is said to precede task T_j if T_j starts after T_i. Similarly, if T_i depends on T_j for its completion, then the relationship is a dependency. The dependency could be in the form of data, where a result of computation produced by one task is used by another (*data dependency*), or it could be a *control dependency* where the execution of one task depends on the execution of another.

In an RT system, tasks are assumed to be periodic. Each task is assumed to be initiated by a *request*, which may arrive from the external world or may be generated internally by the system itself. For example, a task may be assigned the job of reading a number of serial ports, and the ports may generate a request for the task when they are ready. This is an *external request*. On the other hand, the request may be periodically generated internally by an RT timer so that the ports are read with a fixed periodicity. This is an example of an *internal request*.

Every task in a RT system is associated with the following attributes:

Priority—The priority (τ) of a task (T) defines its importance in scheduling. If a set of tasks contends for the CPU, the task with the highest priority is scheduled. The priority may either be *static* (assigned a priori) or *dynamic* (assigned during run time).

Periodicity—The periodicity (P) of a task (T) denotes the time interval at which the task is periodically requested.

Worst-case execution time (WCET)—The WCET (C) of a periodic task (T) is the time required by the task to complete a sequence of actions, once requested, assuming that the task has sole control of the processor and all resources for the task are available.

Response time—The time interval between the instant at which a task is requested and the instant at which the request is served, that is, the specified set of actions in response to the request is completed, is defined as the response time of a task. Clearly, in a multitasking system, the response time of a task may vary depending on states of other tasks and availability of resources. Thus, for a task, the response time is always greater than or equal to its WCET. For a task, T_i, in a set of tasks, with a priority τ_i and with m number of tasks having greater or equal priority, the response time, R_i, can be defined as

$$R_i = C_i + \sum_{k=1}^{m} \left\lceil \frac{R_i}{P_k} \right\rceil C_k \quad (1.1)$$

If a limiting value of the response time exists, the limit is called the *deadline* for the task denoting the instant by which the requested task must be completed and the corresponding interval between the instant at which the task T_i is requested and its deadline is termed as its *relative deadline*, denoted by L_i. A set of periodic tasks $T = [T_1, T_2, \ldots, T_n]$ is schedulable according to a given priority assignment *iff*

$$R_i \leq L_i, \quad \forall i \in [1, n] \quad (1.2)$$

Often the period (P_i) of a task (T_i) is equal to its relative deadline (L_i).

Critical instant—The instant at which request for a task arrives such that the response time of the task is maximum is called its *critical instant*.

Wait time—This is the time interval between the instant at which the request for a task arrives and the instant at which the task is scheduled.

Laxity—The time difference between the WCET and relative deadline of a task is defined as its laxity.

Task overflow—This is a condition where the response time of the task is greater than its deadline.

1.2.3 Scheduling

In a RT system, a task may be in any one of the following states:

Running—The task has control of the CPU.

Ready—The resources for running the task are all available, but a task with a higher priority has control of the CPU.

Blocked—The resources required for running the task are unavailable.

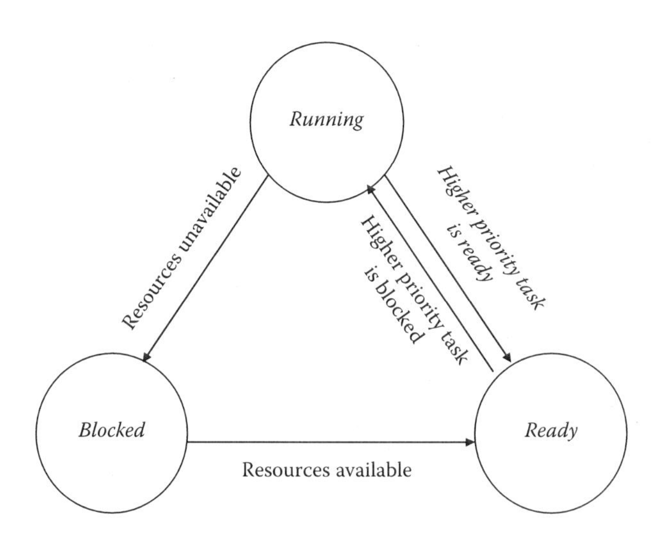

FIGURE 1.2
Task–state transitions in an RT system.

Figure 1.2 shows the possible state transitions associated with a task in an RT system.

The state transitions in an RT system are affected by a *task scheduler*, which is often referred to as a RT scheduler. It is clear that the scheduler is also a process and is a part of the operating system. However, once a task has control of a CPU, it cannot schedule another process—not even the scheduler, if not interrupted. In an RT system, this interrupt is generated by an RT timer and the scheduler is actually an interrupt service routine (ISR) for the timer interrupt and is scheduled periodically at intervals known as the *basic time unit* (BTU), which is often tunable. Since task–state transitions are associated with context switches, context-switch latency must be bounded in an RT system so that deterministic response time is ensured. Clearly, the context-switch latency is a multiple of a BTU. A scheduler maintains a queue of ready processes and the task at the head of the queue is executed by the scheduler. An RT scheduler uses a priority-based preemptive scheduling policy to modify the queue elements.

In order to explain the basic difference between RT and non-RT scheduling, first the following non-RT scheduling policies are introduced, assuming that no dependency or precedence or mutual exclusion relationship exists between any two tasks in the task set $\mathbf{T} = [T_1, T_2, \ldots, T_n]$, and the tasks do not get blocked:

1. Round-robin scheduling
2. First in, first out (FIFO) scheduling
3. Interrupt-driven scheduling

In round-robin scheduling each task T_i $i = 1, 2, \ldots, n$ is allocated a finite quantum of time (C_i) and is scheduled according to a fixed predefined order 1, 2, …, n. Thus, the worst-case wait time is the sum of the time quanta allocated to individual tasks, that is, $\sum_{i=1}^{n} C_i$. FIFO scheduling is similar to round-robin scheduling except that the tasks are scheduled in the order in which they are requested and not in a fixed order. The worst-case wait time remains the same as that of round-robin scheduling. In both cases of scheduling no prioritization is possible.

Interrupt driven scheduling associates tasks that are triggered by external events causing an interrupt to the processor. The corresponding interrupt service routine (ISR) sets a flag in the predesignated position in the scheduling queue. The processor executes the tasks in the order the flags are set in the queue. Thus, if two tasks T_i, T_j associated with two interrupts i, j are requested at the same instant, with *priority* $(i) >$ *priority* (j), then T_i will be executed first followed by T_j. Clearly prioritization is possible in this case. Further, a high-priority task that is frequently requested can cause starvation of other tasks. The worst-case wait time in this case is the maximum of all the task execution times, that is, $\max(C_i)_{\forall i \in [1,n]}$. It is clear from the preceding discussions that the worst-case wait time for a task for all three scheduling policies, and hence the context-switch latency, depends on the application.

As mentioned earlier, an RT scheduler requires a fixed upper bound of context-switch latency and therefore uses a priority-based preemptive scheduling algorithm. In the simplest case, the priority assigned to a task is *static* (assigned a priori and not changed later), *dynamic* (assigned dynamically by the system), or *hybrid* (a combination of both). Examples of static priority-based preemptive scheduling are *rate monotonic* (RM) and *deadline monotonic* (DM) scheduling. *Earliest deadline first* (EDF) and *minimum laxity* scheduling are examples of dynamic scheduling. *Hybrid* scheduling techniques are a combination of RM- and deadline-driven approaches [7]. A comprehensive survey of RT scheduling algorithms with complexity analyses is available in Baruah and Goossens [8].

A static priority assignment algorithm assigns a fixed set of priorities [τ_1, τ_2, …, τ_n] to the set of tasks [T_1, T_2, …, T_n] according to some algorithm. The most common of these is *rate monotonic scheduling* (RMS) [7], which assigns a higher priority to a task having a higher periodicity, and a higher priority task preempts a lower priority task. In this section, different aspects of RMS are visited with the following assumptions:

- No task has a nonpre-emptible section and the cost of preemption is negligible.
- The tasks require more processing than memory and I/O access.
- The tasks are independent and follow no precedence relation.
- All tasks are periodic.

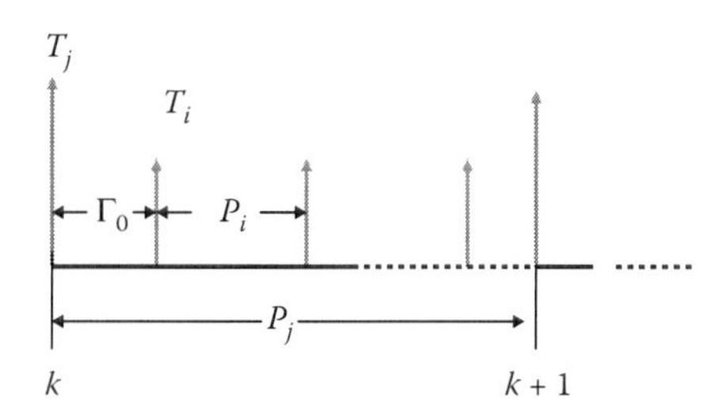

FIGURE 1.3
Timelines of two periodic tasks to establish critical instant of a task.

For an RT system employing priority-based preemptive scheduling, the following important results are proposed in the form of lemmas and proven.

Lemma 1.1

A critical instant for a task occurs when its request arrives along with other higher priority tasks.

Let $\mathbf{T} = [T_1, T_2, \ldots, T_n]$ be a set of tasks with corresponding periods $P_1, P_2, \ldots, P_n$ and WCETs $C_1, C_2, \ldots, C_n$ ordered in a sequence of increasing periods and decreasing priorities from P_1 to P_n.

Figure 1.3 shows the timeline associated with a task T_j, $j \in [1, n]$, which is taken as a reference, and it is assumed that the request for another task T_i, $j > i$, arrives after a time interval Γ_0 counted from the instant at which the kth request for T_j arrives.

It is clear that since T_i has a higher priority than T_j, the response time of task T_j either remains the same or it increases as Γ_0 reduces, since T_j is preempted by T_i at least once when $C_j > \Gamma_0$. Thus, when $\Gamma_0 = 0$, the response time for task T_j is maximum, that is, the critical instant for task T_j occurs when the request for task T_j coincides with that of a higher priority task, T_i. Since, this is true for any two pair of tasks T_i, T_j it follows that it can easily be extended to the entire set of tasks $\mathbf{T}$.

Lemma 1.2

For a set of tasks $\mathbf{T} = [T_1, T_2, \ldots, T_n]$ with periods $P_1, P_2, \ldots, P_n$ and WCETs $C_1, C_2, \ldots, C_n$, the processor utilization can be expressed as $U = \sum_{i=1}^{n}(C_i / P_i)$.

Let $\bar{P}$ denote an interval in which every task receives an integral number of requests such that $\bar{P} = m_i P_i, \forall i \in [1, n]$.

Thus, over the entire interval, the processor utilization denotes the fraction of time utilized by the tasks. This can be written as

$$U = \frac{m_1 C_1 + m_2 C_2 + \cdots + m_n C_n}{\bar{P}} \tag{1.3}$$

Substituting $\bar{P} = m_i P_i, \forall i \in [1, n]$ in Equation 1.3 yields

$$U = \sum_{i=1}^{n} \frac{C_i}{P_i} \tag{1.4}$$

Lemma 1.3

RM scheduling is optimal. If for a set of tasks, there exists a schedule for any other priority assignment, the task set is schedulable by RM scheduling and not vice versa.

Let $\mathbf{T} = [T_1, T_2, \ldots, T_n]$ be a set of tasks with periods $P_1, P_2, \ldots, P_n$ and WCETs $C_1, C_2, \ldots, C_n$.

Let $\mathbf{T}' \subset \mathbf{T}$ be a set of two tasks $[T_i, T_j]$ such that $P_j > P_i$, as shown in Figure 1.4. In Figure 1.4 a case corresponding to the critical instant of the lower priority task is considered to represent the worst case.

If $\mathbf{T}'$ is scheduled using RM scheduling, then T_i should have a higher priority than T_j. If non-RM scheduling is assumed, then for a schedulable $\mathbf{T}'$, a sufficient condition is

$$C_i + C_j \leq P_i \tag{1.5}$$

Now, if $\mathbf{T}'$ is scheduled using RM scheduling and Equation 1.5 holds, then a sufficient condition for $\mathbf{T}'$ to be schedulable is

$$\left\lfloor \frac{P_j}{P_i} \right\rfloor C_i + C_j \leq \left\lfloor \frac{P_j}{P_i} \right\rfloor P_i \tag{1.6}$$

Multiplying both sides of Equation 1.5 with $\lfloor P_j/P_i \rfloor$ yields

$$\left\lfloor \frac{P_j}{P_i} \right\rfloor C_i + \left\lfloor \frac{P_j}{P_i} \right\rfloor C_j \leq \left\lfloor \frac{P_j}{P_i} \right\rfloor P_i \tag{1.7}$$

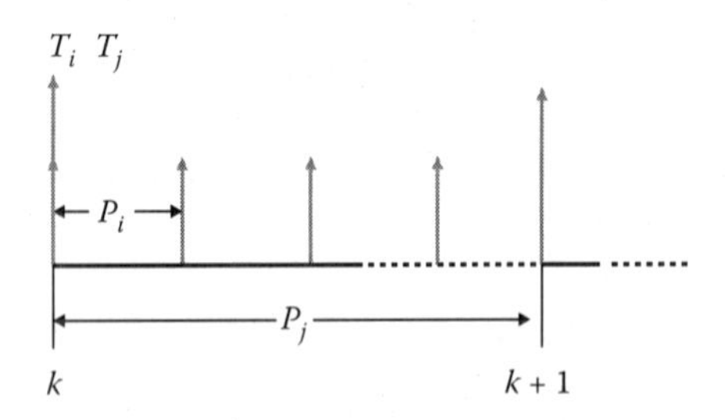

FIGURE 1.4
Timelines of two periodic tasks to prove optimality of RM scheduling.

Since the left-hand side of Equation 1.7 is bigger than the left-hand side of Equation 1.6 and the RHS of both the equations is same, it follows as a natural corollary that if Equation 1.5 holds, then Equation 1.6 holds and not vice versa. Thus, if **T**′ is schedulable by any other scheduling methodology, it is schedulable by RM scheduling and not vice versa. The extension of this lemma from **T**′ to **T** can be done by assuming the task set **T** to be ordered in a sequence of reducing periods.

Lemma 1.4

For a set of two tasks T_1, T_2 with periods P_1, P_2 ($P_2 > P_1$) and WCETS C_1, C_2 schedulable by rate monotonic scheduling, there exists a least upper bound of the maximum value of processor utilization U_{max} given by $U_{max} = 2(\sqrt{2}-1)$.

Since the task set is scheduled by RM scheduling, in the critical zone of T_2 there can be at best $\lceil P_2/P_1 \rceil$ requests for the task T_1 and accordingly, there may be two distinct cases:

Case 1—The $\lceil P_2/P_1 \rceil$-th request of the task T_1 is completed within one request interval of the task T_2, that is, no overlap occurs, as shown in Figure 1.5.

In this case, the following must hold:

$$C_1 \le P_2 - \left\lfloor \frac{P_2}{P_1} \right\rfloor P_1 \tag{1.8}$$

$$C_2 \le P_2 - \left\lceil \frac{P_2}{P_1} \right\rceil C_1 \tag{1.9}$$

Substituting the maximum permissible values for C_1, C_2, the value of maximum processor utilization is obtained as

$$U = 1 + \frac{C_1}{P_1} - \left\lceil \frac{P_2}{P_1} \right\rceil \frac{C_1}{P_2} = 1 + \frac{C_1}{P_2}\left(\frac{P_2}{P_1} - \left\lceil \frac{P_2}{P_1} \right\rceil\right) \tag{1.10}$$

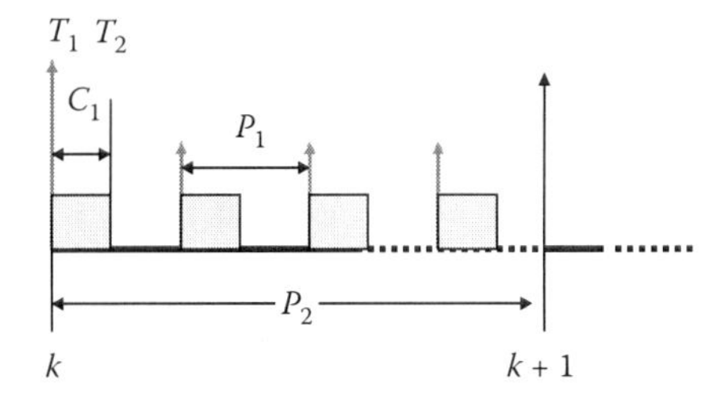

FIGURE 1.5
Timelines of two periodic tasks with no overlap.

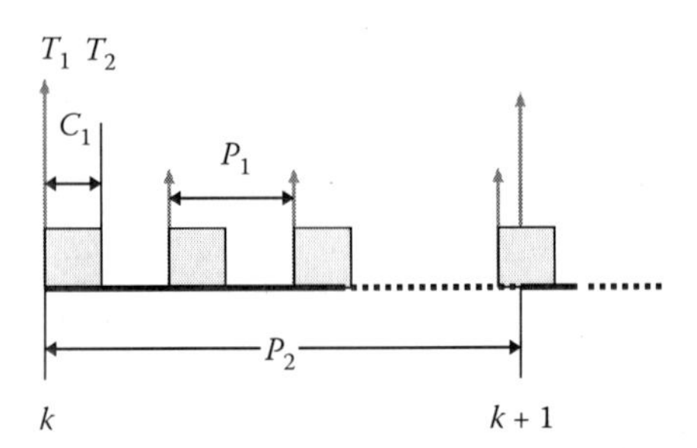

FIGURE 1.6
Timelines of two periodic tasks with overlap.

Note that since $\lceil P_2/P_1 \rceil (C_1/P_2) > (C_1/P_2)$, the processor utilization defined by Equation 1.8 will always be less than 1 and it will reduce with increasing value of C_1.

Case 2—The $\lceil P_2/P_1 \rceil$-th request of the task T_1 is not completed within one request interval of the task T_2 as shown in Figure 1.6.

In this case, the following holds:

$$C_1 \geq P_2 - \left\lfloor \frac{P_2}{P_1} \right\rfloor P_1 \tag{1.11}$$

$$C_2 \leq \left\lfloor \frac{P_2}{P_1} \right\rfloor (P_1 - C_1) \tag{1.12}$$

and in this case, the processor utilization is obtained as

$$U = \left\lfloor \frac{P_2}{P_1} \right\rfloor \frac{P_1}{P_2} + \frac{C_1}{P_1} - \left\lfloor \frac{P_2}{P_1} \right\rfloor \frac{C_1}{P_2} = \left\lfloor \frac{P_2}{P_1} \right\rfloor \frac{P_1}{P_2} + \frac{C_1}{P_2}\left(\frac{P_2}{P_1} - \left\lfloor \frac{P_2}{P_1} \right\rfloor \right) \tag{1.13}$$

It is clearly seen that in this case U increases with increasing C_1 from a minimum defined in Equation 1.11 subject to a maximum satisfying the inequality 1.12 with the constraint $C_2 \geq 0$.

An examination of Equations 1.8 and 1.11 shows that there exists a lower bound of maximum processor utilization and this occurs when

$$C_1 = P_2 - \left\lfloor \frac{P_2}{P_1} \right\rfloor P_1 \tag{1.14}$$

Substituting the value of C_1 from Equation 1.8 in either Equation 1.10 or Equation 1.13 yields the minimum upper bound of maximum processor utilization U_{max} as

$$U_{\max} = 1 - \frac{(1-x)x}{x + \lfloor P_2 / P_1 \rfloor} \tag{1.15}$$

where

$$x = \frac{P_2}{P_1} - \left\lfloor \frac{P_2}{P_1} \right\rfloor \tag{1.16}$$

From Equation 1.15 it is seen that the minimum value of $U_{\max}$ occurs when $\lfloor P_2/P_1 \rfloor = 1$. Substituting $\lfloor P_2/P_1 \rfloor = 1$ in Equation 1.15 and solving for $(dU_{\max}/dx) = 0$ yields

$$x = -1 \pm \sqrt{2} \tag{1.17}$$

Discarding the extraneous value $x = -1 - \sqrt{2}$ and substituting $x = -1 + \sqrt{2}$ in Equation 1.15 yields

$$U_{\max} = 2\left(\sqrt{2} - 1\right) \tag{1.18}$$

Lemma 1.5

For a set of tasks schedulable by rate monotonic scheduling, there exists a least upper bound of the maximum value of processor utilization U given by $U = n(2^{1/n} - 1)$.

Let $\mathbf{T} = [\Gamma_1, \Gamma_2, \ldots, \Gamma_n]$ be a set of tasks with periods $P_1, P_2, \ldots, P_n$ and WCETs $C_1, C_2, \ldots, C_n$, ordered in increasing sequence of request periods, that is, $P_1 < P_2 \cdots < P_n$. It is further assumed that

$$\left\lfloor \frac{P_i}{P_j} \right\rfloor \neq \left\lceil \frac{P_i}{P_j} \right\rceil \quad \forall i, j \in [1, n] \tag{1.19}$$

and

$$\left\lceil \frac{P_i}{P_j} \right\rceil \leq 2 \quad \forall i \in [1, n] \tag{1.20}$$

Equation 1.19 states that the ratio of the request periods of any two tasks is a fraction, and Equation 1.20 states that within the request period of any task, there can be at best two requests for any other task. This is a standard assumption considered by contemporary researchers [7] and the lemma can only be proven under these assumptions.

Since the task set is schedulable by RM scheduling, the following holds:

$$\begin{aligned}
&\left\lceil \frac{P_2}{P_1} \right\rceil C_1 + C_2 \le P_2 \\
&\left\lceil \frac{P_3}{P_1} \right\rceil C_1 + \left\lceil \frac{P_3}{P_2} \right\rceil C_2 + C_3 \le P_3 \\
&\ldots \\
&\left\lceil \frac{P_n}{P_1} \right\rceil C_1 + \left\lceil \frac{P_n}{P_2} \right\rceil C_2 + \cdots + C_n \le P_n
\end{aligned} \tag{1.21}$$

Applying the constraint (Equation 1.20) in Equation 1.21 yields

$$\begin{aligned}
&2C_1 + C_2 \le P_2 \\
&2C_1 + 2C_2 + C_3 \le P_3 \\
&\vdots \\
&2C_1 + 2C_2 + \cdots + C_n \le P_n
\end{aligned} \tag{1.22}$$

This further yields

$$\begin{aligned}
&C_1 = P_2 - P_1 \\
&C_2 = P_3 - P_2 \\
&\vdots \\
&C_{n-1} = P_n - P_{n-1} \\
&C_n = P_n - 2(C_1 + C_2 + \cdots + C_{n-1})
\end{aligned} \tag{1.23}$$

Let

$$g_i = \frac{P_n - P_i}{P_i}, \quad i = 1,2,3\ldots n \tag{1.24}$$

or, $g_i P_i = P_n - P_i$.

Thus

$$C_i = P_{i+1} - P_i = g_i P_i - g_{i+1} P_{i+1} \tag{1.25}$$

Again, substituting the value of C_i from Equations 1.24 and 1.25 in Equation 1.23 yields

$$C_n = P_n - 2g_i P_i, \quad i = 1, 2, \ldots, n-1 \tag{1.26}$$

Now rewriting $U = \Sigma_{i=1}^{n}(C_i/P_i) = \Sigma_{i=1}^{n-1}(C_i/P_i) + (C_n/P_n)$ yields

$$U = \sum_{i=1}^{n-1} \frac{g_i P_i - g_{i+1} P_{i+1}}{P_i} + \frac{P_n - 2g_i P_i}{P_n}$$

or

$$U = \sum_{i=1}^{n-1}\left(g_i - g_{i+1}\frac{P_{i+1}}{P_i}\right) + \left(1 - 2g_i\frac{P_i}{P_n}\right)$$

or

$$U = 1 + g_1\left(\frac{g_1 - 1}{g_1 + 1}\right) + \sum_{i=2}^{n-1} g_i\left(\frac{g_i - g_{i-1}}{g_i + 1}\right) \tag{1.27}$$

Now, the upper bound of processor utilization reaches a minimum value when

$$\frac{\partial U}{\partial g_i} = 0 \quad i = 1, 2, \ldots, n-1$$

Defining $g_0 = 1$ produces

$$\begin{aligned}
\frac{\partial U}{\partial g_1} &= \frac{(g_1^2 + 2g_1 - 1)}{(g_1 + 1)^2} - \frac{g_2}{g_2 + 1} \\
\frac{\partial U}{\partial g_2} &= \frac{(g_2^2 + 2g_2 - g_1)}{(g_2 + 1)^2} - \frac{g_3}{g_3 + 1} \\
&\vdots \\
\frac{\partial U}{\partial g_{n-1}} &= \frac{(g_{n-1}^2 + 2g_{n-1} - g_{n-2})}{(g_{n-1} + 1)^2} - \frac{g_n}{g_n + 1}
\end{aligned} \tag{1.28}$$

Equating each $\partial U/\partial g_i$ term to 0 in Equation 1.28 yields

$$
\begin{aligned}
g_1 + 1 &= \sqrt{2(g_2 + 1)} \\
g_2 + 1 &= \sqrt{2(g_3 + 1)} \\
&\vdots \\
g_{n-1} + 1 &= \sqrt{2(g_n + 1)}
\end{aligned}
\tag{1.29}
$$

In general

$$g_i = 2^{(n-i)/n} - 1 \tag{1.30}$$

Substituting these values in Equation 1.25 yields

$$U_{\max} = n(2^{1/n} - 1) \tag{1.31}$$

As a direct corollary of Lemma 1.5 it follows that as the number of tasks n increases, the limiting value of $U_{\max}$ is obtained as

$$U_{\max_\infty} = \lim_{n \to \infty} n(2^{1/2} - 1) \tag{1.32}$$

which yields

$$U_{\max_\infty} = 0.693 \tag{1.33}$$

It follows from Equation 1.32, therefore, a set of tasks is always schedulable by RM scheduling if the maximum processor utilization of the task set is less than 0.693. This is very conservative, but the condition is both necessary and sufficient.

In order to better explore whether the constraint on $U_{\max}$ defined by Equation 1.32 can be relaxed, the concept of *demand bound function* $D_f(\Delta t)$ is introduced, which defines the maximum time allocated to the tasks in any time interval Δt. A set of tasks **T** is schedulable *iff* $D_f(\Delta t) \leq \Delta t$. This is possible with dynamic scheduling algorithms like the EDF. This is a dynamic scheduling algorithm that schedules the task with the earliest deadline. Like RM scheduling, EDF is also optimal; if a task set is schedulable EDF can schedule it too. EDF may be preemptive as well as nonpreemptive.

Figure 1.7 illustrates EDF with a set of three tasks: T_1, T_2, T_3. Each task T_i is assumed to be associated with a tuple $\{C_i, L_i\}$. For the case shown in Figure 1.7, the set of tuples for the task set $[T_1, T_2, T_3]$ is $[\{4, 1\}, \{5, 2\}, \{8, 2\}]$ and as

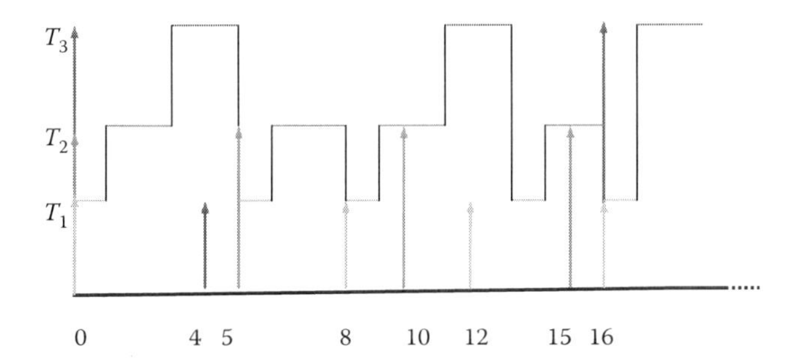

FIGURE 1.7
(**See color insert.**) A representative execution profile with EDF.

illustrated, the processor utilization is unity, that is, $U = 1$ signifying that the processor never stays idle.

Now, if it can be assumed that $P_i = L_i$ for each task T_i in the task set represented in Figure 1.7 and RM scheduling is applied, then the execution profile obtained in Figure 1.8 is obtained.

With RM scheduling, applying $n = 3$ in Equation 1.32 yields $U_{\max} = 3(\sqrt[3]{2}-1)$, which is equal to 0.77. But the corresponding value of the actual processor utilization for the task set considered is 0.9. Thus, RM scheduling does not guarantee a no-overflow condition for the task set considered.

As mentioned earlier, embedded systems are a class of RT systems that are often portable in nature with constraints on power consumption. For such systems processors with dynamic voltage scaling (DVS) are often very useful. A DVS processor operates in multiple modes in which the processor speed and hence power consumption is modulated depending upon the load requirement. While deadlines and periodicity of some tasks may remain unaltered, WCETs of individual tasks in a task set can be altered so that power consumption is minimized and yet the task set remains schedulable. This can be done by *assigning* the tasks in a task set to specific DVS modes.

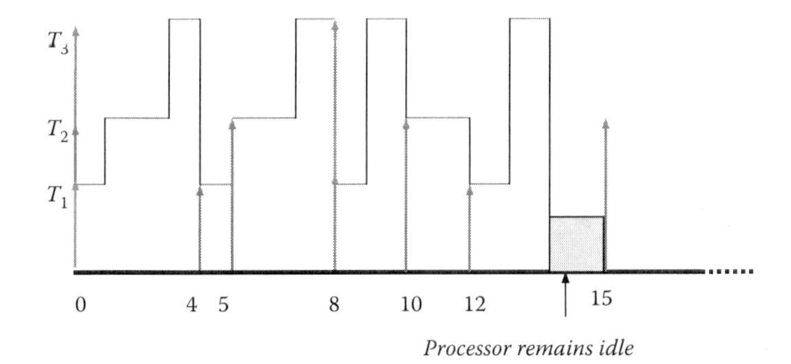

FIGURE 1.8
(**See color insert.**) Execution profile with RM scheduling corresponding to Figure 1.7.

A technique for scheduling RT tasks on a DVS processor has been proposed in Reference 9.

In order to reduce the context-switch latency in an RT system, many systems implement tasks as *threads*. A process (program in execution) may be associated with multiple threads. A thread is a *lightweight* process implying that all threads belonging to a process share the same address space with all data, code, and status, which results in a reduced context-switch latency. A thread inherits the priority of the parent process and some RTOSs allow modification of thread priorities further. It must also be noted that individual processes in an RT system use separate stacks.

1.2.4 Synchronization between Tasks

A real-time system is essentially a multitasking system and task synchronization is essential for the tasks to share common resources like memory buffers and I/O devices or for sequencing of tasks when there exists a specific precedence or dependency relationship. An RTOS normally provides two types of synchronization: *explicit synchronization* and *implicit synchronization.*

1.2.4.1 Explicit Synchronization

Explicit synchronization is a type of synchronization in which the task IDs are known to the synchronized task entities. The synchronization is achieved using a technique called *cross stimulation.* Figure 1.9 illustrates explicit synchronization.

The advantage of this kind of synchronization is the speed. The task IDs need to be known to all the participating tasks and this is usually done by an initialization task. However, Figure 1.9 has been specifically chosen to represent explicit synchronization to bring out a case where this should not be used. In Figure 1.9 the task T_1 periodically stimulates task T_2 and thus there exists a strong cohesion between the two tasks, and since T_2 succeeds

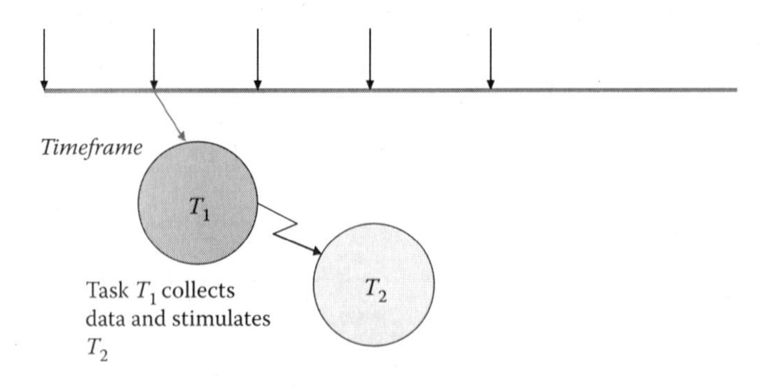

FIGURE 1.9
Explicit synchronization.

T_1 a contention for shared resources shall never arise in case the tasks constitute a producer–consumer pair. On the contrary, if T_1 is a timer process that generates a periodic wakeup for T_2, then this makes sense. Thus explicit synchronization is more useful for precise task sequencing rather than managing sharing of resources.

1.2.4.2 Implicit Synchronization

Implicit synchronization is always achieved using the RTOS, and the participating processes do not need to use each other's task IDs. Figure 1.10 represents the different types of implicit synchronization mechanisms for both single processor and multiprocessor systems.

1.2.4.2.1 Flag Semaphores

The use of semaphores for handling shared resources in a GPOS is well known. An RTOS uses semaphores in a similar way and implements several variants to take care of specific RT requirements. As in a GPOS, a *set* operation sets a flag semaphore and a *reset* resets it. To use a shared resource, a task has to first set the semaphore. A task that tries to set a semaphore in the set state or a task that tries to reset a semaphore in the reset state gets blocked.

To understand the use of a flag semaphore, the producer–consumer system represented in Figure 1.9 is revisited and this time a flag semaphore is used to synchronize the producer and the consumer. The necessity of synchronization becomes clear if one considers a case where the producer writes multiple data elements all pertaining to a time instant on a buffer, and the consumer reads them periodically. At the kth instant, the data belonging to the $(k-1)$th instant remaining in the buffer is overwritten with fresh data by

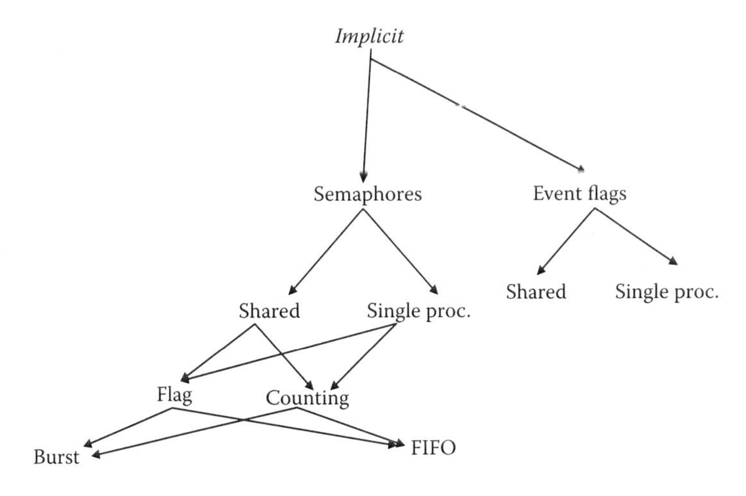

FIGURE 1.10
Implicit synchronization.

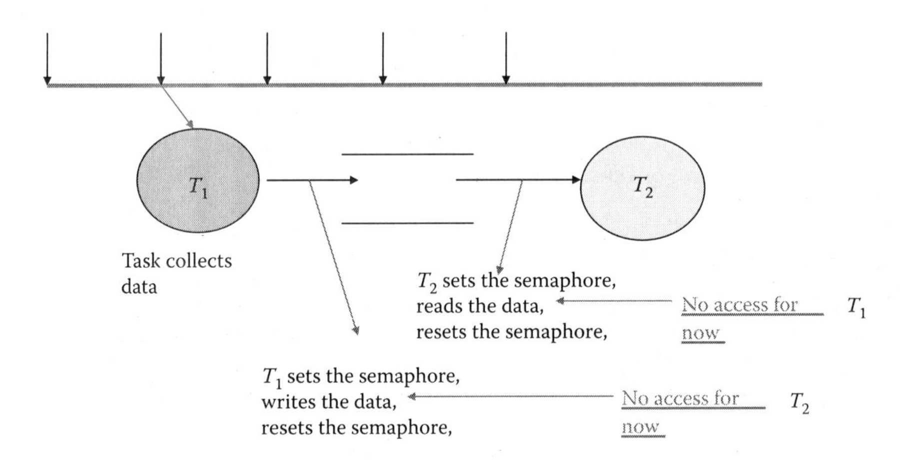

FIGURE 1.11
Synchronization using a flag semaphore.

the producer. Now, if it so happens that the producer is preempted when it has written half the data elements and the consumer starts reading before the write operation is complete, then the temporal significance of the data will not be preserved and the system may behave erroneously. A flag semaphore can be used to synchronize the producer and the consumer as shown in Figure 1.11.

The difference between the case represented in Figure 1.9 and the one represented in Figure 1.11 is that in the former case task T_2 succeeds task T_1 and in the latter case task T_2 is dependent on task T_1. Thus, in this case, the tasks T_1 and T_2 may have asynchronous timelines with one task blocking the other only while accessing the buffer.

There are, however, some problems with flag semaphores, the first and foremost being the *deadly embrace*. Figure 1.12 illustrates this for a set of two tasks T_1, T_2 synchronized using two semaphores S_1, S_2.

It is clearly seen that in this case both the tasks will be perpetually in the blocked state resulting in a *deadlock* in the following scenario considering that the task T_2 has a higher priority compared to T_1; T_1 sets the semaphore S_1 and gets preempted by T_2; T_2 now sets the semaphore S_2 and gets blocked while trying to set semaphore S_1; and T_1 is now scheduled and gets blocked while trying to set semaphore S_2. One way of preventing this is the introduction of a *critical section* in the lower priority task such that it cannot be preempted within the critical section.

Next, extending this simple synchronization mechanism to a case involving multiple producers and a single consumer is attempted. It is first assumed that the *m* producers each write on a different buffer and the consumer reads these buffers. Since there are *m* shared resources, the synchronization between the producers and the consumer shall require *m* flag semaphores. Further, at the consumer end, the consumer can get blocked

```
Pseudocode for task T1
While(forever)
{
set semaphore S1;
set semaphore S2;
---------
reset semaphore S1;
reset semaphore S2;

}
```

```
Pseudocode for task T2
While(forever)
{
set semaphore S2;
set semaphore S1;
---------
reset semaphore S2;
reset semaphore S1;

}
```

FIGURE 1.12
Deadlock problem with flag semaphores.

m times within its request period to read data from *m* producers—in the worst case—sequentially, that is, and in that case if any of the producers get blocked after setting the corresponding semaphore, the consumer may get blocked perpetually. Figure 1.13 represents the pseudocode for the *i*th producer and the consumer.

At the consumer end, this causes a problem. The consumer tries to set the semaphores sequentially and it might so happen that one or more of the producers may get preempted after setting the corresponding semaphore, which will block the consumer perpetually.

```
Pseudocode for ith producer

While(forever)
{
set semaphore si;
write_data();
reset semaphore si;

}
```

```
Pseudocode for the consumer

While(forever)
{
for(i=0;i<m;i++)
{
set semaphore si;
read_data();
reset semaphore si;
}
}
```

FIGURE 1.13
Problem with multiple semaphores.

1.2.4.2.2 Counting Semaphores

An alternative scheme for handling multiple synchronizing entities and multiple synchronized resources without use of multiple flag semaphores is the use of a counting semaphore. A counting semaphore is associated with an integer count with a maximum value (*MAX_COUNT*) and two operations, namely, *increment* and *decrement*. The increment is equivalent to a set operation for a flag semaphore and a decrement is equivalent to a reset. A counting semaphore is initialized during startup and every increment increases the count by one. Similarly, every decrement reduces the count by one. A process that tries to increment the counting semaphore once the count reaches MAX_COUNT blocks. Similarly, a process that tries to reset the semaphore once the count reaches the initial value (usually 0) blocks.

If now, one tries to examine the synchronization between the producers and the consumers, it is seen that the synchronization can be reliably achieved using a single counting semaphore with *MAX_COUNT* initialized to the number of producers. Each producer will try to increment the counting semaphore before writing the data on the buffer, and the consumer will decrement the semaphore each time it reads the buffer. A producer will block if there are no free slots in the buffer (count equals MAX_COUNT) and will be unblocked once the consumer decrements it. In this way, it is easy to ensure that the consumers do not cause the buffer to be overwritten, and the consumer is also prevented from reading old data. If p denotes the current count value of the semaphore, the consumer must be programmed to read to location p, while each producer must be programmed to write at the location $p + 1$.

1.2.4.2.3 Event Flags

Now, this simple scheme of synchronization using counting semaphore, as described earlier, is very basic. It does not allow the consumer to know exactly which producer has written the data unless the data is labeled. Moreover, if one tries to synchronize the control flow of the producer and the consumer so that the consumer reads the entire buffer after all or certain specific combinations of producers have written on the buffer. To achieve this, some RTOSs allow a distinct synchronization mechanism called an *event flag*. An event flag can be viewed as a byte or a word, each bit of which represents an event. A process may be programmed to be able to access a shared resource when a condition represented by a logical combination of the bits is satisfied.

Figure 1.14 illustrates how this can be used to synchronize the set of producers and consumers under consideration.

Thus, one way could be the ith producer setting the bit D_i and the consumer getting access to the resource when the Boolean condition $\Pi_i^m D_i = 1$ is satisfied for an m producer system.

At a first glance an event flag may look like a logical combination of multiple flag semaphores. However, the difference becomes clear when one

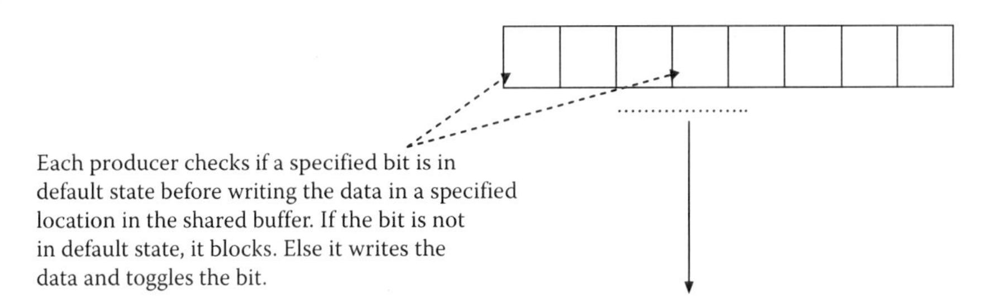

FIGURE 1.14
Synchronization using event flags.

considers that the consumer becomes blocked only when it tries to access the shared resource and the predefined Boolean condition is not satisfied; it does not get blocked on individual flag semaphores sequentially. The advantage over counting semaphores is also clear when one considers the ease with which the data can be associated with a producer without explicit tagging. This advantage becomes clearer if one assumes a condition where one or more producers have different periodicity, a condition that can be handled by using an appropriate Boolean function.

1.2.4.2.4 Burst and FIFO Mode Semaphores

Some RTOSs allow flag and counting semaphores to operate in either a *burst* mode or an *FIFO* mode. In burst mode, when several processes wait on a semaphore, tasks get unblocked in decreasing sequence of their priorities, that is, the process with the highest priority gets unblocked first, flowed by the next, and so on. In case of an FIFO mode semaphore, the processes get unblocked in the sequence in which they got blocked, that is, the process entering the queue first gets unblocked first. The execution profile for a four-task system with burst and FIFO mode semaphores are illustrated with Figures 1.15 and 1.16, respectively.

It is clear from Figure 1.16 that the response time of higher priority tasks is lower with burst mode semaphores.

1.2.4.2.5 Priority Inversion

Priority inversion is a typical situation where a medium priority task preempts a lower priority task while it has access to a shared resource for which a higher priority task is waiting. The situation is explained in Figure 1.17.

In Figure 1.17, it is assumed that the priority sequence of the tasks is $T_1 > T_2 > T_3$. The task T_1 sets a semaphore and accesses a shared resource when it is preempted by the task T_1, which runs for some time and gets

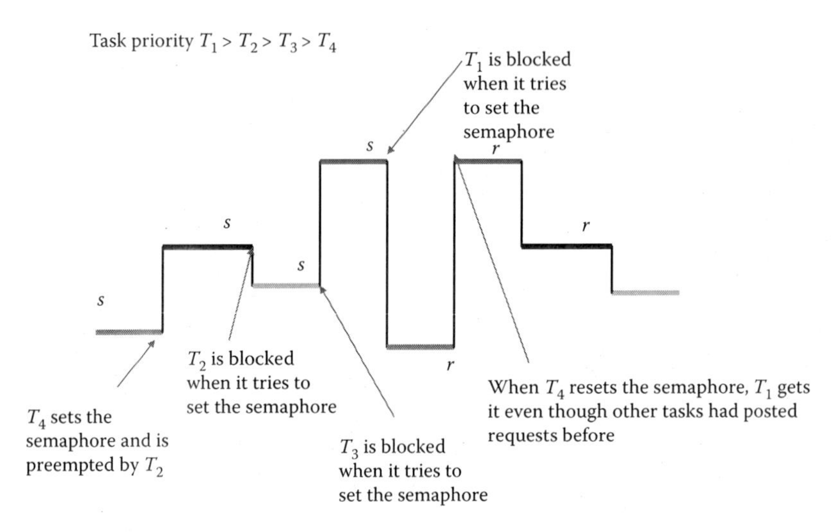

FIGURE 1.15
(**See color insert.**) Burst mode semaphore.

blocked while trying to access the same shared resource. Now, even though T_1 is otherwise ready, it will be displaced by a lower priority task T_2 (if it is ready) and a priority inversion is said to have occurred.

Priority inversion can lead to dangerous situations in mission critical applications. The most famous recorded instance of priority inversion is associated with the NASA Mars Pathfinder mission [10]. The problem occurred

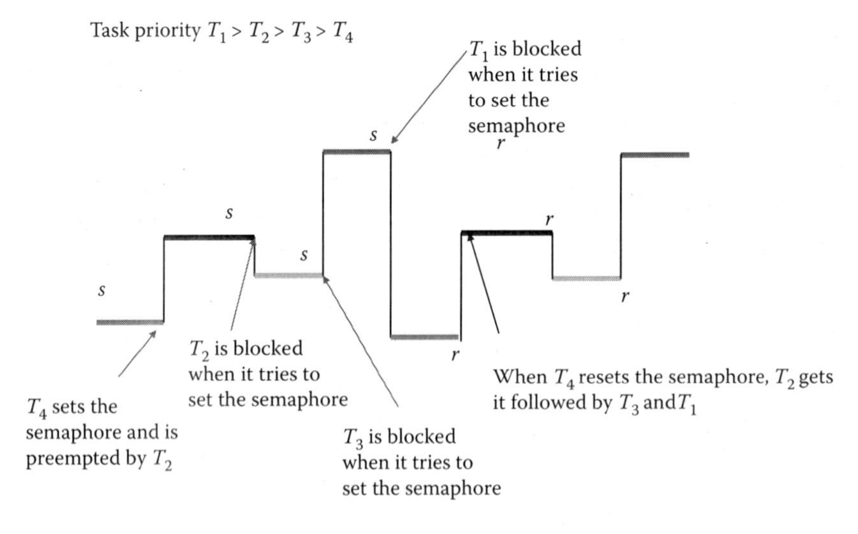

FIGURE 1.16
(**See color insert.**) FIFO mode semaphore.

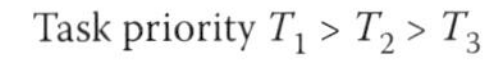

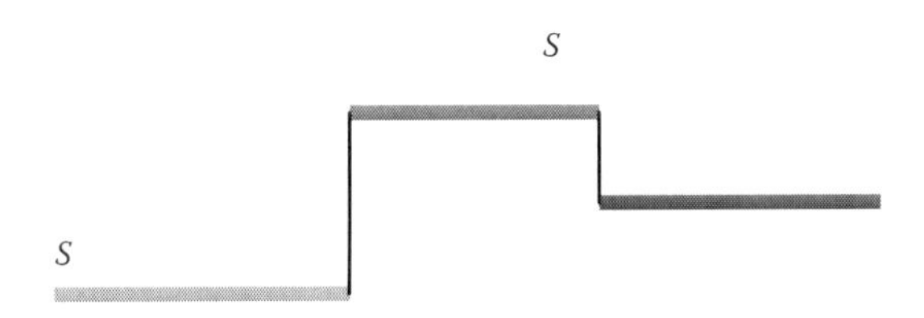

FIGURE 1.17
(See color insert.) Execution profile under priority inversion.

on the Mars sojourner. The system used two high-priority tasks, which controlled the activity over the data bus through which the devices communicated. One of the tasks was a lower priority meteorological data collection task that received data from one of the high priority bus manager tasks and was preempted by medium priority tasks accessing the communication resource. The high priority bus manager, thus, got blocked and the medium priority tasks continued to run. The situation was detected by a watchdog process that led to a series of system resets.

The problem of priority inversion can be solved by a number of workarounds, namely, *priority inheritance*, *priority association*, and the use of *critical sections*. Priority inheritance elevates the priority of the lower priority task to a level more than that of the medium priority task while it is accessing the resource. That is to say, for the case shown in Figure 1.17 the priority of T_3 must be raised to a level that is more than that of T_2 while it accesses the shared resource so that T_1 can resume once T_3 releases the resource. Priority association, on the other hand, is a resource-specific strategy that associates a priority level to a resource and this is equal to the priority level of its highest priority contender plus one. When a task accesses this resource, it inherits the priority of the resource and is thus prevented from preemption. Introduction of a *critical section* in the lowest priority task while it accesses the shared resource is the third work around. A critical section defines a portion of the code where the executing task cannot be preempted. Thus, for the case illustrated in Figure 1.17 the task T_3 may access the shared resource within a critical section and will not be preempted before releasing it, preventing priority inversion.

In the computation of the least upper bound of maximum processor utilization with RM scheduling, it has been assumed that the tasks in a given task set are independent and do not interact. However, tasks do interact and while a critical section in a lower priority task extends the execution time of a higher priority task by a finite quantum of time, a priority inversion can cause an unbounded extension. Thus, if task T_i is ready but is blocked for a finite quantum of time B_i by a lower priority task entering a critical section, for example, then its utilization may be assumed to be extended [11] by an amount B_i/T_i and thus the set of n tasks remain schedulable by RM scheduling if

$$\sum_{i=1}^{n} \frac{(C_i + B_i)}{P_i} \leq n(2^{1/n} - 1) \tag{1.34}$$

Again, in a practical RT system, there are sporadic tasks, for example, special recovery routines that are activated by exceptions. Augmentation of Equation 1.34 to handle sporadic tasks is based on a conservative approach by assigning the highest periodicity to such a task in the computation of maximum processor utilization [12].

1.2.5 Timer Functions in a Real-Time Operating System (RTOS)

Many RTOSs provide timer functions available as RT system calls to the applications programmer. Typical timer functionalities provided are

Delay function—Delays the execution of a task by a specified delay interval.

Calendar clock execution—Specifies execution of a task at a specific time instant.

Periodic execution—Specifies execution of a task periodically with a specified interval.

The implementation and syntax of the calls are RTOS specific and are usually accessible to the application programmer through an application programming interface (API).

1.2.6 Intertask Communication in an RTOS

Most RTOSs provide standard intertask communication structures for data transfer, such as shared variables, bounded buffers, message queues, mail boxes, and FIFO. The concepts are similar to those for a GPOS.

1.3 Virtual Deadlock

In most real-time systems, shared resources, particularly I/O devices, are handled in a mutually exclusive manner. That is to say a higher priority task contending for a shared resource is not allowed to preempt a lower priority task while it is accessing the same resource. Note, this could have also been a workaround for the priority inversion problem illustrated in Figure 1.17, but is not adopted in order to preclude a *virtual deadlock* condition.

The virtual deadlock can be understood by examining a situation where two tasks T_1 and T_2 try to access a common I/O device like an external firmware, which requires a finite sequence of commands. A typical case could be where none of the tasks get to complete initialization of the I/O device as each task is forced to relinquish the shared resource by the other before completion of the initialization. Such a situation may lead to a virtual deadlock.

1.4 Scheduling in a Distributed Real-Time System Environment

Scheduling in a distributed real-time system (DRTS) environment is more complicated that scheduling in an RTOS running on a single processor because of two reasons: scheduling in a multiprocessor environment and message interchange between tasks that introduce dependency. If $\mathbf{T} = [T_1, T_2, \ldots, T_n]$ is a set of tasks with periods $P_1, P_2, \ldots, P_n$ and WCETs $C_1, C_2, \ldots, C_n$ scheduled over m processors, then a necessary condition for the set of tasks to be schedulable is

$$\mu = \sum_{i=1}^{n} \frac{C_i}{P_i} \leq m \tag{1.35}$$

With messages involved, task scheduling has to take into account message scheduling and this is either *dynamic* or *static*. It is difficult to guarantee tight deadlines by dynamic scheduling techniques in a uniprocessor system if mutual exclusion and task precedence constraints are considered.

Static scheduling in a DRTS environment, on the other hand, involves computation of a *trace* of tasks and messages before execution, which is stored for dispatch during run time. It is assumed that all clocks are globally synchronized and dispatchers across all processors have a common time reference. The trace is defined as the *schedule* and the quantum of time after which the schedule repeats is called the *schedule period*. One way of computation of a static schedule is *task and message coscheduling*, which assumes a definite transmission time for each message, and a CPU time for sending and receiving messages, and a nonpreemptive use of the shared network connecting the processors. Figure 1.18 explains static coscheduling in a DRTS with three tasks (T_1, T_2, T_3) assigned to three processors (p_1, p_2, p_3) interconnected over a network with message dependencies M_{13} and M_{21}, assuming that the periods of the tasks P_1, P_2, P_3 each equal to the schedule period P_s. Again, it is assumed that the relative deadline of each task is its period.

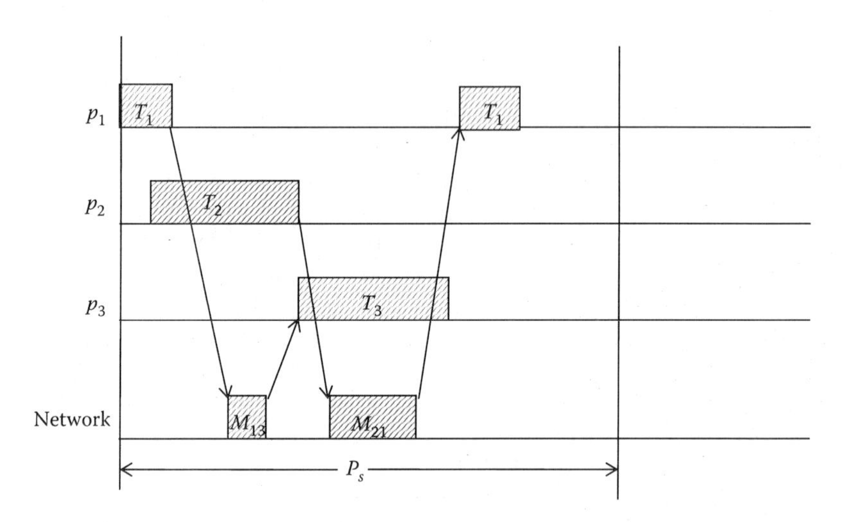

FIGURE 1.18
(**See color insert.**) Coscheduling in DRTS.

However, it might be possible to meet the dependency and deadline requirements of the tasks by an alternate schedule, as shown in Figure 1.19.

Computing a static schedule on a distributed environment is known to be a NP-complete task [13], and for small task sets often a heuristic search is applied using a search tree. A search tree starts with an empty schedule with the start node of the precedence graph for a given task set describing the dependencies associated with transmitted messages. Each level of the search tree corresponds to an instant when a task and/or message is scheduled. Outward edges from a node in the search tree point to tasks/messages

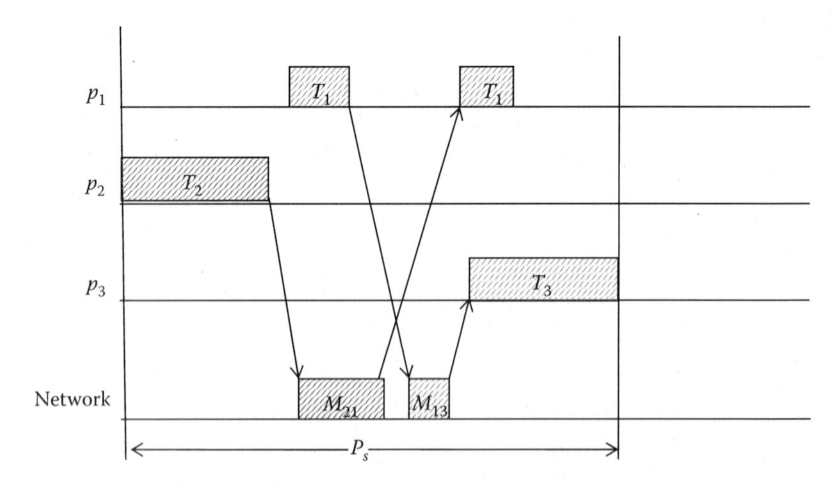

FIGURE 1.19
(**See color insert.**) An alternate schedule.

that can be scheduled next, fulfilling the dependency constraints. A path from the root node to a particular node at level l records the sequence of scheduling decisions that have to be made up to level l. The search completes when the set of unscheduled tasks and messages is empty. A path from the root to the leaf of the search tree describes a complete schedule. A complete schedule that meets the deadline is said to be feasible. As an example, schedules shown in Figures 1.18 and 1.19 are both feasible for the task set considered. Branch and bound heuristics [13] is often employed to find a feasible schedule with the shortest time. An examination of Figures 1.18 and 1.19 shows that the schedule shown in Figure 1.18 is shorter than the schedule shown in Figure 1.19. For a general case where the task periods are unequal, the schedule period is the least common multiple (LCM) of the individual task periods.

NUMERICAL AND ANALYTICAL PROBLEMS

1.1. An RT system consists of four scanner tasks that scan sensor data from the field in a real-life plant and write on to a buffer that is read by a data handler once every 100 ms. Each scanner takes 5 ms and the data handler takes a further 20 ms. If the whole system is controlled by a heartbeat timer that takes 20 ms each time it is triggered, what should be the maximum periodicity of the timer, so that no overflow occurs? Assume RM scheduling for the tasks.

1.2. In Problem 1.1, if it is assumed that one of the scanners updates the data at a higher periodicity, that is, once in 50 ms and the data handling load for this increases by 40%, what should be the new period for the heartbeat timer?

1.3. An RT embedded system is developed around an Intel 8088-based system with a single 8259. It comprises three tasks—T_1, T_2, T_3—with priorities in the same sequence, which are connected to IRQs 4, 5, and 6, respectively, with a default priority assignment for the IRQs. If the tasks are associated with execution times of 20, 25, and 30 ms, respectively, and the interrupt latency is 5 μs, what is the response time of the system, if interrupt driven scheduling is assumed? If the period at which the triggers arrive is the same and it is P, what should be the minimum value of P if no overflow occurs, assuming scheduling overhead is 10%? Sketch the execution profile for two cycles with this value of P.

1.4. Three tasks—T_1, T_2, T_3—with priorities in the same sequence are synchronized using a flag semaphore. Initially, T_1 and T_2 are in the blocked state when T_3 runs for 50 ms after having set the semaphore, when it is preempted by T_2. T_2 runs for 30 ms more before it is blocked again, while trying to set the same semaphore, and T_3 runs for a further period of 10 ms when it is preempted by T_1, which runs for 20 ms and gets blocked again when it tries to set the semaphore set by T_1. T_3 then runs for 15 ms and resets the semaphore. If T_2 and T_1 take further 20 ms each to complete the activities for the particular cycle, draw the execution profile and calculate the time spent by T_1 in blocked mode assuming (a) a burst mode semaphore and (b) an FIFO mode semaphore. What is the processor utilization in this case? Does this depend on the execution profile or the semaphore type?

1.5. For Problem 1.2 derive a suitable synchronization mechanism for the application and explain your answer with appropriate pseudocodes.

1.6. Investigate if an EDF can schedule a set of tasks with the set of $\{C_i, L_i\}$ tuples defined as [{3, 1}, {5, 2}, {8, 3}] and state what happens if an overflow occurs.

References

1. MATLAB. Homepage. www.mathworks.com.
2. Wind River. VxWorks RTOS. www.windriver.com/products/vxworks/.
3. Microsoft. Welcome to Windows CE 5.0. msdn.microsoft.com/en-us/library/ms905511.aspx.
4. Enea. Enea OSE. www.enea.com/solutions/rtos/ose.
5. OSEK VDX Portal. www.osek-vdx.org/.
6. TinyOS. www.tinyos.net/.
7. Liu, C. L., and Layland, J. W. Scheduling algorithms for multi-programming in a hard real-time environment. *Journal of the ACM*, 20(1), 46–61, 1973.
8. Baruah, S., and Goossens, J. Real-time scheduling: Algorithms and complexity. http://www.ulb.ac.be/di/ssd/goossens/baruahGoossens2003–3.pdf.
9. Seo, E. Energy efficient scheduling of real-time tasks on multicore processors. *IEEE Transactions on Parallel and Distributed Computing*, 19(11), 1540–1552, 2008.
10. Di Natale, M. An introduction to real-time operating systems and schedulability analysis. http://inst.eecs.berkeley.edu/~ee249/fa07/RTOS_Sched.pdf.
11. Zalewski, J. What every engineer needs to know about rate monotonic scheduling: A tutorial. http://www.academia.edu/3086988/What_Every_Engineer_Needs_to_Know_about_Rate-Monotonic_Scheduling_A_Tutorial.
12. Kopetz, H. *Real-Time Systems: Design Principles for Distributed Embedded Applications*, 2nd ed. Springer, 2011.
13. Abdelzaher, T. F., and Shin, K. G. Combined task and message scheduling in distributed real-time systems. *IEEE Transactions on Parallel and Distributed Systems*, 10(11), 1179–1191, 1999.

2

Distributed Real-Time Systems

In Chapter 1, the enabling concepts behind real-time systems were introduced. In this chapter, the concepts that extend a real-time system to a distributed real-time system (DRTS) are introduced.

2.1 Architectures

Classically, a DRTS is a collection of several real-time subsystems interconnected using a shared data network. Each subsystem, in the simplest case, could be a *processing element* comprising a processor and memory. The entire DRTS may be viewed as a loosely coupled distributed system with definite response time, that is, real-time functionality. Architectural issues of a DRTS mostly focus on the interconnection of the constituent subsystems and also the ways the functionalities of a DRTS are governed. First, the interconnection topologies are visited.

2.1.1 Topologies of Interconnection

A *topology* defines how the components of a DRTS are connected. The interconnection may be *physical* or *logical*—two nodes may be physically connected but they may never communicate. The basic physical topologies widely used in a DRTS are *bus*, *star*, and a *ring* as shown in Figure 2.1.

The bus topology consists of a single network cable on which the individual nodes are connected by shorter cables. This is most suitable for systems where the subsystems are arranged along a line; a typical example could be an assembly line. A star topology, on the other hand, is the most common in DRTS applications. The switch acts as an interconnection providing a path between any two nodes. Logically, this is a strongly connected graph with an edge existing between any two nodes. On the contrary, a ring topology provides physical segmentation of the network.

2.1.2 Time- and Event-Triggered Architectures

While topology is an important architectural issue in a DRTS, the architecture could either be *event-triggered* or *time-triggered*. An *event* is a change in

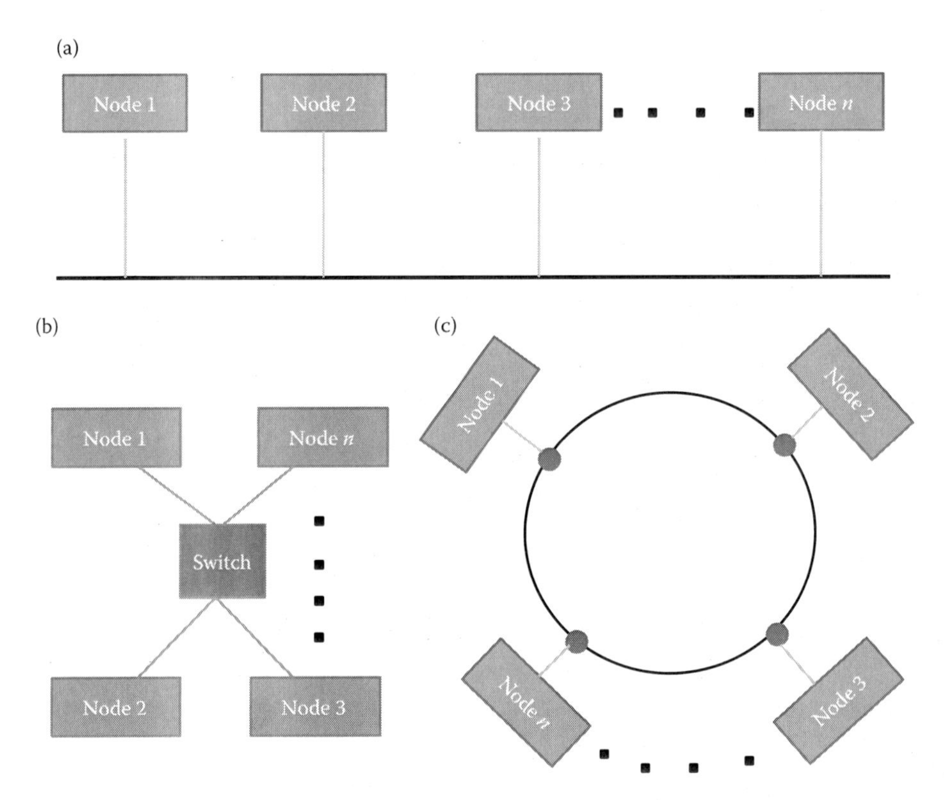

FIGURE 2.1
Topologies of interconnection in a DRTS. (a) Bus, (b) star, (c) ring.

state of a system. For example, a clock tick generated by an oscillator may be viewed as a *timer event*. If the actions of a system are controlled by a timer event, the system is time-triggered. On the contrary, examples of event-triggered control could be activation of an interrupt service routine in a microprocessor. Events may be synchronous with the control flow of a system; they are called *internal events*. Or, they may be generated externally, and they are called *external events*. Event-triggered distributed systems involve transmission of *state messages* between the processing elements. The state of the DRTS is updated based on the state messages. Proper sequencing is maintained by time-stamping of events that could either be in *transmission sequence ordering* (TSO) when they are ordered in the sequence the messages are transmitted by a sender or *receive sequence ordering* (RSO) in which they are ordered in the sequence they are received by a receiver. Whereas time-triggered architectures of DRTS ensure a high predictability, event-triggered architectures allow a high flexibility.

2.2 Properties

The main properties of a DRTS [1] may be enumerated as (a) *modularity*, (b) *flexibility*, (c) *composability*, and (d) *dependability*. Figure 2.2 is used to understand these properties.

Figure 2.2 shows a centralized scheme of automotive control where all major systems in an automobile are controlled by a single dashboard computer. The primary drawback of such a system is clear from Figure 2.2 itself, namely, failure of the dashboard computer may cause a complete system shutdown. Moreover, the system lacks flexibility. For example, if the transmission control module has to be modified, it will not be possible to make alterations on the live system, as the tasks are all run on a single computer. An obvious solution, therefore, appears to be a decentralized system, as shown in Figure 2.3.

However, a question arises: Is it sufficient to distribute the functionalities of the composite system among different processors? The answer is no. For a DRTS, on the other hand, composability rather than divisibility is more important. *Composability* is defined as the property of a system that allows a composite system to have a desired set of functionalities defined in terms of the functionalities of the individual constituent subsystems. This requires the rules governing the interconnection between two component subsystems to be defined. The architecture of a DRTS should be such that the interconnect rules are supported. For example, if the DRTS shown in Figure 2.3

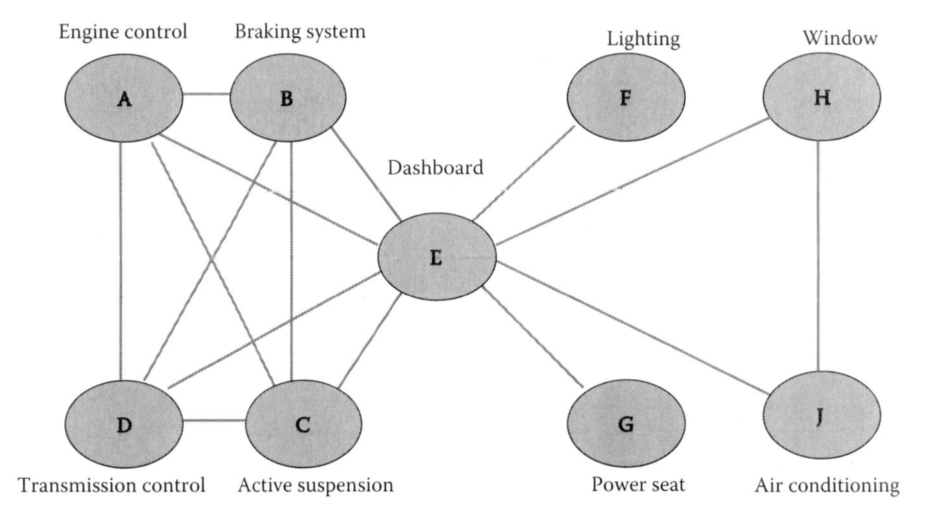

FIGURE 2.2
(See color insert.) Automotive control-centralized scheme.

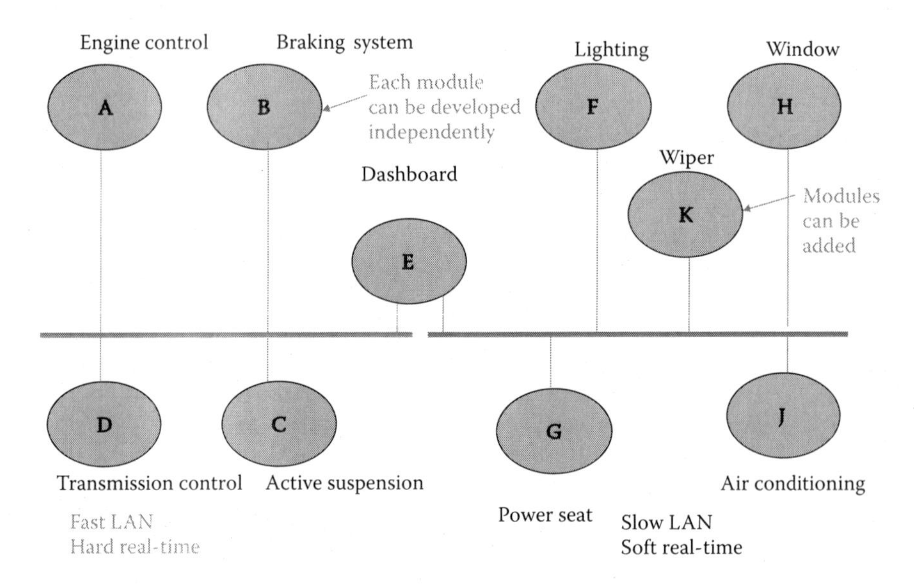

FIGURE 2.3
(**See color insert.**) Modularity and flexibility in a DRTS.

is used for controlling an electric vehicle, the interface between the engine control and the transmission control should be defined so that the composite system can deliver sufficient torque while moving uphill. For a DRTS, *dependability* refers to the degree of trust with which the system is believed to be able to deliver a set of services. This depends on several attributes like reliability, maintainability, and availability.

2.3 Time

Time is perhaps the most important concept in DRTS, for both event-triggered and time-triggered architectures. A *timeline* can be defined as an infinite progression of time, and an *instant* is a cut on the timeline. An event is associated with a particular instant. Mathematically, a timeline is an infinite set of instants denoted by {**T**}. The order of two events e_1,e_2 associated with two instants a,b in the timeline is defined as the *temporal order*. The interval $[a,b]$ on a timeline is defined as a *time interval*. If the timeline is such that it is always possible to find an instant c on the timeline between a,b, however small $[a,b]$ might be, then the timeline is termed as *dense*. Else the timeline is *sparse*. A sparse timeline is classically an infinite progression of instants, each separated from the preceding one by a fixed time interval called the *granularity* of the timeline. Most real-time systems operate on a sparse timeline, which is generated by a real-time clock that produces *microticks* at a fixed granularity to update a counter.

Since a DRTS comprises multiple real-time subsystems that operate on a single or multiple timelines, there needs to be a reference or a standard. Some universal standards of time are the International Atomic Time (TAI), which defines a second as 9,192,631,770 cycles of the standard Cs-133 transition [2]; and the Coordinated Universal Time (UTC) [2], which defines time in terms of period of rotation of the Earth. Accordingly, a clock in DRTS could be a *physical clock* for which the counter indicates *physical time*. In many DRTS applications, it is often sufficient to establish a temporal order in a sequence of events generated by the component subsystems, each operating with a *logical clock* and a single *global clock* in the resultant ensemble, which generates a reference timeline or the *timeframe*. In the rest of the discussion that follows, the term *actual* is used to refer to the timeframe.

In an ensemble of N logical clocks, each logical clock generates a timeline constituted by a progression of microticks, $Z_i(k)$, being elapsed time recorded by the ith logical clock, $i \in [1,2, \ldots, N]$, at the instant defined by the kth microtick and is referred to as time at kth microtick for the sake of simplicity. If $i = r$ represents the reference clock, $Z_r(k)$ is taken as the actual value of elapsed time from the start at the instant defined by the kth microtick. In the present context, the progression of k is defined with respect to the timeframe and $Z_i(k)$ is obtained by a cut on the timeline of the ith logical clock or the ith timeline at the kth microtick of the timeframe. If it be such that the timeframe is synchronized to a physical clock, then $Z_r(k)$ indicates a physical time.

Usually the local clocks use an oscillator and the frequency of the oscillators change with time, typically in the tune of 10^{-3} to 10^{-7} s. Since the clocks drift, the time interval $\Delta T_i(k) = Z_i(k+1) - Z_i(k)$ may be different from the reference value $\Delta T_r(k) = Z_r(k+1) - Z_r(k)$ on the timeframe and the drift rate for the local clock i at the kth microtick denoted by $\rho_i(k)$ may be defined as

$$\rho_i(k) = \frac{\Delta T_r(k) - \Delta T_i(k)}{\Delta T_r(k)} \tag{2.1}$$

In Equation 2.1, ΔT_r denotes the actual time interval between two successive microticks of any clock, that is, $Z_r(k+1) - Z_r(k)$. ΔT_r is then defined as the granularity of local clocks denoted by g, assumed to be the same for all clocks, in the present context:

$$g = \Delta T_r \tag{2.2}$$

Accordingly, the offset $\sigma_{i,j}(k)$ between two local clocks i,j at the kth microtick is defined as

$$\sigma_{i,j}(k) = Z_i(k) - Z_j(k) \tag{2.3}$$

and in an ensemble of N clocks, the *precision* is defined as

$$\delta = \max_k(\sigma_{ij}(k)),\ i,j \in [1,2,\ldots,N] \tag{2.4}$$

The offset of a local clock with respect to the global clock is defined as its *accuracy*.

In a DRTS, n consecutive microticks are grouped to form a *macrotick* or a *macrogranule* G and this is the interval at which the time stamp generated by a clock is incremented. Thus,

$$G = ng \tag{2.5}$$

and all events are time-stamped with granularity G. An event e generated at the kth microtick of the local clock i will have a time stamp

$$t_i(e) = \left\lfloor \frac{Z_i(k)}{G} \right\rfloor \tag{2.6}$$

The drift in the local clocks may cause events generated at the same instant on the timeframe to have two different time stamps at two subsystems, as shown in Figure 2.4 where the time-stamp of an event e generated by the ith clock is one more than the time stamp generated by the jth clock because of the clock drift. In the rest of the discussions, it is further assumed that if an event e occurs at the kth microtick, then $Z_i(e) = Z_i(k)$.

In a DRTS, it is often very important to reconstruct the temporal order of events from the time stamps. Reconstruction of a temporal order of a set of events means reconstruction of the sequence of events from their actual time stamps. Since events are time-stamped according to a Send order(by the sender) or Receive Order(by the receiver), they always carry the time stamp put on by a local clock and there exists an uncertainty in the time stamp put by the local clock with respect to the actual time stamp.

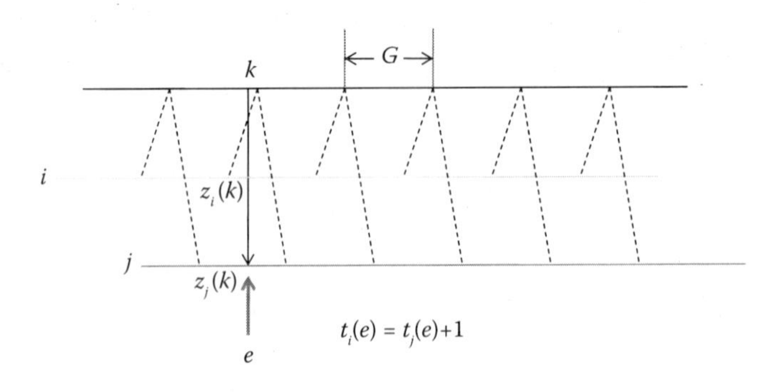

FIGURE 2.4
Timeframe and local clocks.

In order to ascertain this uncertainty, it is first assumed that the timeframe is *reasonable*. A timeframe is said to be reasonable if the following inequality holds:

$$G > \delta \tag{2.7}$$

It can be shown, for a reasonable timeframe, that the following Lemma holds.

Lemma 2.1

If the timeframe is reasonable, then for any two local clocks i,j in an ensemble of n local clocks, the respective time stamps for an event e can at most differ by unity, that is, if $G > \delta$, then $|t_i(e) - t_j(e)| \leqslant 1$.

Since, $G > \delta$,

$$\frac{|Z_i(k) - Z_j(k)|}{G} < 1 \tag{2.8}$$

Now substituting

$$\frac{Z_i(k)}{G} = \left\lfloor \frac{Z_i(k)}{G} \right\rfloor + f_i, f_i \in \mathbb{R}^+ \tag{2.9}$$

and

$$\frac{Z_j(k)}{G} = \left\lfloor \frac{Z_j(k)}{G} \right\rfloor + f_j, f_j \in \mathbb{R}^+ \tag{2.10}$$

in Equation 2.8 yields

$$|t_i(e) - t_j(e) + f_i - f_j| < 1 \tag{2.11}$$

In Equations 2.9 and 2.10 and throughout the rest of the book, $\mathbb{R}^+$ is assumed to represent the set of positive real numbers including 0.

Since $t_i(e), t_j(e) \in \mathbb{Z}^+$, where $\mathbb{Z}^+$ is the set of all positive integers including 0, it follows easily that $|t_i(e) - t_j(e)| \in \mathbb{Z}^+$.

From Equation 2.11, it follows therefore,

$$|t_i(e) - t_j(e)| \in \left[\left\lfloor 1 - |f_i - f_j|_{\max} \right\rfloor, \left\lfloor 1 + |f_i - f_j|_{\max} \right\rfloor\right] \tag{2.12}$$

However, since $f_i, f_j \in [0,1)$, it follows from Equation 2.12

$$|t_i(e) - t_j(e)| \in [0,1] \tag{2.13}$$

Hence,

$$|t_i(e)-t_j(e)|\leq 1 \quad \text{if } G>\delta \tag{2.14}$$

Lemma 2.1 is used to state and prove Theorems 2.1 and 2.2.

Theorem 2.1

The maximum uncertainty in ascertaining the actual time stamp $t(e)$ of an event e, from its time stamp generated by the local clock i, that is, $t_i(e)$, is ±1 if the timeframe is reasonable.

The proof follows easily from Lemma 2.1.

Since i,j are chosen arbitrarily from the ensemble, one of them can be taken as a reference, and then the following holds

$$|t_i(e)-t_r(e)|\leq 1 \quad \forall i\in[1,2,\ldots,N], i\neq r \tag{2.15}$$

Thus in the extreme case,

$$t_i(e)=t_r(e)\pm 1 \tag{2.16}$$

Equation 2.16 proves the theorem.

Theorem 2.2

The maximum uncertainty in ascertaining the actual duration of a time interval between the instants at which any two events e_1,e_2 occur from the time stamps $t_i(e_1),t_j(e_2)$, respectively, is $\pm 2G$ for a reasonable timeframe.

From Equation 2.16 it follows in the extreme case,

$$t_i(e_1)=t_r(e_1)\pm 1 \tag{2.17}$$

$$t_j(e_2)=t_r(e_2)\pm 1 \tag{2.18}$$

$$|t_r(e_2)-t_r(e_1)|=|t_j(e_2)-t_i(e_1)|\mp 2 \tag{2.19}$$

Again, using Equations 2.6 and 2.19, it follows that in the extreme case,

$$|Z_r(e_2)-Z_r(e_1)|=|Z_j(e_2)-Z_i(e_1)|\mp 2G \tag{2.20}$$

The proof follows from Equation 2.20. It must be remembered that the result presented in Equation 2.20 assumes constant offset during a macrotick. A more general proof of this can be derived using interval arithmetic.

2.3.1 Reconstruction of Temporal Order for Burst Periodic Events: A Sufficiency Condition

In a DRTS, let a set of periodic events $\mathbf{E} = [e_1, e_2, \ldots, e_n]$ recur with a period Δ.

If e_{pq} $\forall p \in [1,2, \ldots, n]$, $q \in \mathbb{Z}^+$, $q > 0$ denotes the qth *instance* (or occurrence) of an event $e_p \in \mathbf{E}$, then the following holds:

$$Z_r(e_{p(q+1)}) - Z_r(e_{pq}) = \Delta \tag{2.21}$$

Since the events are burst periodic, it is assumed further that

$$|Z_r(e_{pq}) - Z_r(e_{sq})| \leq \pi \quad \forall e_q, e_s \in \mathbf{E}, q \neq s \tag{2.22}$$

that is, for a particular instance, all events in $\mathbf{E}$ occur within a time interval π constituting a burst of events.

Now, if $\pi \leqslant \delta$ holds, then the actual time stamps $t_r(e_{pq})$ for the qth instance shall be same for all events $e_p \in \mathbf{E}$, and the corresponding time stamp generated by an arbitrary ith clock in the ensemble shall differ by ± 1, that is,

$$t_i(e_{pq}) = t_r(e_{pq}) \pm 1 \tag{2.23}$$

and

$$t_i(e_{p(q+1)}) = t_r(e_{p(q+1)}) \pm 1 \tag{2.24}$$

If a causal relationship exists between the qth instance of an event e_p and another event e_s ($e_p, e_s \in \mathbf{E}$), then the following must hold:

$$t_i(e_{s(q+1)}) - t_i(e_{pq}) > 1 \tag{2.25}$$

Again, since $t_r(e_{pq}) = t_r(e_{sq})$, Equation 2.25 implies

$$Z_i(e_{p(q+1)}) - Z_i(e_{pq}) \geq G \tag{2.26}$$

and the sufficient condition for which is

$$Z_r(e_{p(q+1)}) - Z_r(e_{pq}) - 2G \geq G \tag{2.27}$$

or

$$Z_r(e_{p(q+1)}) - Z_r(e_{pq}) \geq 3G \tag{2.28}$$

By definition, the left-hand side of Equation 2.28 is Δ, and, thus, it can be stated that if a burst of events $\mathbf{E}$ occurring within a time span π recurs after a time period Δ, then the DRTS is termed the π/Δ precedent. From Equation

2.28 it is seen that the temporal order of events belonging to two consecutive instances of **E** can always be constructed by any arbitrary clock i in the ensemble if the burst repeats after a period $\Delta = 3G$ and the system is said to be $0/3G$ precedent.

The significance of the π/Δ precedence can be understood if one considers a practical case like a networked control system (NCS) where a number of process control loops are integrated over a shared network. Each control loop consists of a sensor that senses the process variables periodically with an interval T_s and sends the sensor data to a controller that generates a controller output with the last two samples of sensor data. If all the sensors transmit at the same time, then the transmission events constitute a burst. Since the controller has to establish a precedence among two successive samples of sensor data, $T_s \geqslant \Delta$ must hold.

2.4 Time Synchronization

Time synchronization in a DRTS is a process by which the clocks of all component subsystems are adjusted so that the difference in time lies within a bound. The time interval at which time synchronization is repeated is termed the *resynchronization interval* Γ. The main requirements of time synchronization in a DRTS are as follows:

1. The synchronization algorithm must ensure that a predefined precision is always achieved at any point of time.
2. When temporal order has to be established using physical time, the global clock should be synchronized to a physical clock.
3. The synchronization algorithm should be fault tolerant, that is, capable of handling processor and communication faults.
4. The synchronization algorithm should not be drastic. That is to say, the adjustment should not be such that the time becomes negative at any instant on a timeline.
5. The synchronization algorithm should be scalable.
6. The synchronization algorithm should preferably be light, that is, it should not degrade the performance of a DRTS.

Accordingly, synchronization may either be *external*, where the global clock is synchronized to a physical clock, or *internal*, where logical clocks are synchronized.

In a DRTS, since time synchronization requires a finite time, for example, the time taken to read other clocks and the time taken to adjust a clock, there elapses a finite time interval between the instant time synchronization is initiated and the time at which it is completed, however small it might be. This is particularly important for a DRTS where the component subsystems are integrated over a network where the time taken to read other clocks may vary. Thus, there exists a finite difference between the clocks even after they are synchronized and the smallest difference by which two clocks can be synchronized is defined as *convergence* ν. Now, if ρ_{max} is the maximum drift rate of a clock during the resynchronization interval Γ, then two clocks may drift by a maximum interval

$$\phi = 2\Gamma\rho_{max} \tag{2.29}$$

In order for a precision δ to be maintained, the following must hold:

$$\phi + \nu \leq \delta \tag{2.30}$$

Equations 2.29 and 2.30 may be used to ascertain the resynchronization interval in a DRTS.

2.4.1 External Clock Synchronization

External clock synchronization algorithms are mostly *centralized* using a central master or decentralized. A few of these are discussed next.

2.4.1.1 Centralized External Clock Synchronization: Cristian's Algorithm

Cristian's algorithm for external clock synchronization was proposed by Flaviu Cristian in his seminal paper [3] and is essentially a *centralized pull* algorithm where clocks are synchronized to a physical clock acting as the *master* and the other clocks are *slaves*. Figure 2.5 illustrates Cristian's algorithm.

At regular intervals of $(\delta/2\rho_{max})$ seconds, each slave sends a request to the time server to send the physical time T. If T_R be the time instant at which the request is sent, θ is the time interval after which the server sends the physical

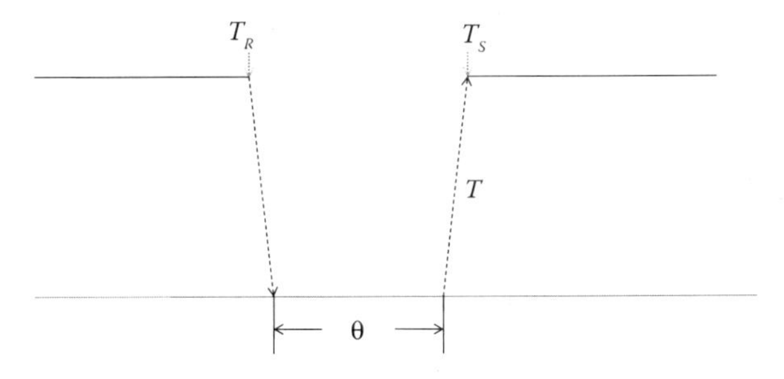

FIGURE 2.5
External clock synchronization: Cristian's algorithm.

time T, and T_S be the time instant at which the slave receives the communication from the time server, then

$$L = \frac{T_s - T_R - \theta}{2} \tag{2.31}$$

The slave corrects the value of T by adding L to it. Several estimates of L are used to obtain a realistic estimate, since communication latencies are variable.

2.4.2 Centralized Clock Synchronization Algorithm: Berkeley Algorithm

The Berkeley algorithm is an averaging scheme based on a single master. The master polls every slave for time value and on receipt of the same computes an average discarding the outliers. The correct value is sent to all slaves for adjustment of local clocks.

2.4.3 Decentralized Clock Synchronization: Network Time Protocol (NTP)

Network Time Protocol (NTP) uses a hierarchy of multilevel servers exchanging time information for synchronization. Figure 2.6 illustrates this between two levels of servers $i,i+1$.

The server at level i communicates a time value T_1 to the server at level $i+1$. If the offset between the two timelines is δ and the latency of communication is L and T_2 is the instant at which the server at level $i+1$ receives this communication, then the following holds from Figure 2.6:

$$T_2 = T_1 + \delta + L \tag{2.32}$$

$$T_3 = T_2 + \theta \tag{2.33}$$

$$T_4 = T_3 - \delta + L' \tag{2.34}$$

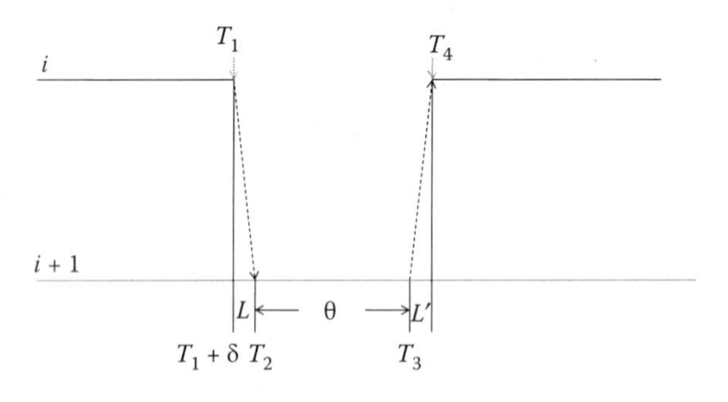

FIGURE 2.6
External clock synchronization: Network Time Protocol (NTP).

Substituting the value of T_1, T_2 from Equations 2.18 and 2.19 in Equation 2.20 yields

$$T_4 = T_1 + \theta + L' + L \quad (2.35)$$

which is independent of the offset between the timelines of the two servers.

The term $L + L'$ represents the round-trip message transmission latency between the two servers on the network. If the interval θ is known, then the accuracy of synchronization depends on the value of the round-trip message passing latency only.

2.4.4 Byzantine Clocks and Fault-Tolerant Clock Synchronization

A Byzantine clock is a *two-faced* clock, that is, a clock that reads differently when read by two clocks at the same instant. The problem of time synchronization with a two-faced clock was termed the Byzantine clock problem by Lamport, Shostak, and Pease [4]. Thereafter, a series of papers from Leslie Lamport have established a class of algorithms for fault tolerant synchronization algorithms that can handle Byzantine faults and these have come to be known as *Lamport clock synchronization* algorithms [4–6].

If it is assumed that every clock reads every other clock at a resynchronization instant and applies a simple average of all clock values, then such a synchronization is bound to fail under if a Byzantine clock places its clock values in *symmetrical* positions. This is explained as follows with two healthy clocks i,j and a Byzantine clock f. If it be assumed that $\delta = 2\mu$ is the precision of the ensemble, then a clock will discover no anomaly if the other clock values lie within a band $\pm\mu$ with respect to its value. Defining $Z_{if}(k)$ and $Z_{jf}(k)$ as the values of time associated with the clock f when read by the healthy clocks i and j respectively, then the case shown in Figure 2.7 may arise.

From Figure 2.7 it is seen that clocks i and j work on the sets of time values $[Z_{if}(k), Z_i(k), Z_j(k)]$ and $[Z_{jf}(k), Z_i(k), Z_j(k)]$, respectively, which both satisfy the precision, but applying a simple average cannot bring the clocks i,j any closer.

Lamport clock synchronization algorithms can be classified into two classes: *interactive convergence* and *interactive consistency*. These algorithms are reported by Lamport and Melliar-Smith [6], and the reader is encouraged to refer to this paper for an in-depth analysis of these. These algorithms are capable of achieving synchronization in an ensemble of N clocks with

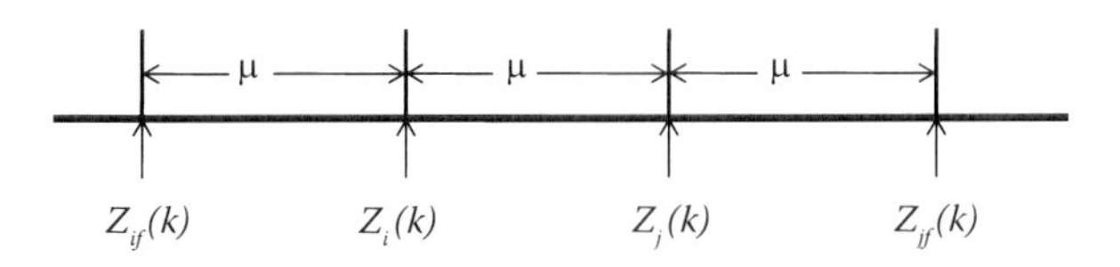

FIGURE 2.7
Byzantine clocks with time values in symmetrical positions.

M Byzantine clocks if $N \geqslant 3M+1$. In the following sections, first the interactive convergence algorithm is discussed followed by the interactive consistency algorithm. The implicit assumption behind these algorithms is that each clock in the ensemble can read every other clock.

2.4.4.1 Interactive Convergence Algorithm

The interactive convergence algorithm is explained as follows. At any arbitrary instant k, if a clock i reads clock $j(i \neq j)$, then the difference Δ_{ij} is calculated as follows:

$$\Delta_{ij}(k) = \begin{matrix} Z_i(k) - Z_j(k), & \text{if } |Z_i(k) - Z_j(k)| \leq \mu \\ 0 & \text{if } |Z_i(k) - Z_j(k)| > \mu \end{matrix} \tag{2.36}$$

If $Z_{ij}(k)$ denotes the time recorded by clock j as estimated by clock i, then

$$Z_{ij}(k) = Z_i(k) + \Delta_{ij}(k) \tag{2.37}$$

Noting that $\Delta_{ii}(k) = 0$, the corrected value of clock i after synchronization shall be

$$Z_i^c(k) = Z_i(k) + \frac{\sum \Delta_{ij}(k)}{N} \forall i, \quad j \in [1, 2, \ldots, N] \tag{2.38}$$

In Equation 2.38, the term $(\sum\Delta_{ij}(k)/N)$ $\forall i,j \in [1,2, \ldots, N]$ denotes the average value of the sum of the differences between any two clocks in the ensemble (including self, which is assumed to be 0) computed using Equation 2.36. It can be proved that the interactive convergence algorithm is capable of achieving an effective synchronization in an ensemble of N clocks with M Byzantine clocks if $N \geqslant 3M+1$. For this purpose, Theorem 2.3 is proposed and proved.

Theorem 2.3

In an ensemble of N clocks consisting of M Byzantine clocks, the interactive convergence algorithm is capable of achieving synchronization with a precision $\delta = 2\mu$ if $N \geqslant 3M+1$.

From Figure 2.7 it is seen that for each Byzantine clock f, the worst case occurs when a Byzantine clock r is such that

$$|\Delta_{if}(k) - \Delta_{jf}(k)| = 3\mu \tag{2.39}$$

where i,j are two healthy clocks, the maximum difference in the average value of any two clocks i,j for each Byzantine clock will differ by 3μ.

Again, since there are M Byzantine clocks, applying maximum difference in the average term computed by any two clocks i,j in the ensemble is

$$\Delta_{\max} = \frac{3M\mu}{N} \tag{2.40}$$

Again, for the synchronization to be effective,

$$\frac{3M\mu}{N} < \mu \tag{2.41}$$

or

$$N \geq 3M + 1 \tag{2.42}$$

which proves Theorem 2.3.

It is to be noted that the treatment of the interactive convergence algorithm in this section assumes that all clocks are read at the same instant. This may not be the case. However, since each clock computes the difference with other clocks, the reading instants may be different. The convergence of the clock values is iterative and spans several resynchronization intervals.

2.4.4.2 Interactive Consistency Algorithm

Unlike the interactive convergence algorithm that uses a simple average, an interactive consistency algorithm uses a median to eliminate the outliers. Lamport and Melliar-Smith [6] discuss two algorithms, one of which is capable of achieving synchronization in an ensemble of N clocks with M Byzantine clocks such that $N \geqslant 3M + 1$ and another with fewer, that is, $N \geqslant 2M + 1$ clocks. In this section, the basic algorithm (i.e., the first) is discussed, and the reader is encouraged to refer to Lamport and Melliar-Smith for the second.

The consistency algorithm presented here is derived from the basic *Byzantine general* problem described by Lamport et al. [4]. The basic OM(M) algorithm [4] is modified to the COM(M) interactive consistency algorithm, which is explained here in the time synchronization parlance with $M = 1$ and $N = 4$, and the algorithm is as follows:

1. Each clock is assumed to be associated with a unique processor. Each processor $i \in [1,2,3,4]$ retains its clock value $Z_i(k)$ to each of the other three processors.
2. For a particular i, each processor $j \in [1,2,3,4]$, $j \neq i$ that receives the value $Z_i(k)$ relays the difference $\Delta_{ij}(k) = Z_i(k) - Z_j(k)$ to each of the other two processors.

$$1: \ [Z_1(k)] \ [Z_2(k), Z_{32}(k), Z_{42}(k)] \ [Z_3(k), Z_{23}(k), Z_{43}(k)] \ [Z_4(k), Z_{24}(k), Z_{34}(k)]$$

$$2: \ [Z_2(k)] \ [Z_3(k), Z_{43}(k), Z_{13}(k)] \ [Z_4(k), Z_{14}(k), Z_{34}(k)] \ [Z_1(k), Z_{41}(k), Z_{31}(k)]$$

$$3: \ [Z_3(k)] \ [Z_4(k), Z_{14}(k), Z_{24}(k)] \ [Z_1(k), Z_{21}(k), Z_{41}(k)] \ [Z_2(k), Z_{12}(k), Z_{42}(k)]$$

$$4: \ [Z_4(k)] \ [Z_1(k), Z_{21}(k), Z_{31}(k)] \ [Z_2(k), Z_{32}(k), Z_{12}(k)] \ [Z_3(k), Z_{13}(k), Z_{23}(k)]$$

FIGURE 2.8
Clock values and estimates at different processors for COM(1) algorithm.

Thus, at the end of these two steps, defining $Z_{ij}(k)$ as the estimate of the jth clock by the ith processor as defined by Equation 2.37, the clock values and estimates available with the four processors are represented in Figure 2.8.

It is seen that the clock values and estimates at each processor i consists of four sets **A**,**B**,**C**,**D**, of which the set **A** is the value of its own clock and the remaining sets **B**,**C**,**D** represent the clock value of another processor j and the estimate of the clock j by the remaining two processors l,m. Since only one clock can be faulty, effective synchronization can be achieved by first computing the medians of each of the sets of **A**,**B**,**C**,**D** and then computing the median of the resultant values.

As in the case of the interactive convergence algorithm using difference values allows reading of clocks at different instants. One drawback of the COM(M) algorithm is its high message complexity, which is $O(N^M)$.

2.4.5 Probabilistic Clock Synchronization

Most of the aforementioned algorithms work with bounded message passing latencies. It has been shown by Landelius and Lynch [7] that in an ensemble of N clocks, if each clock runs in real-time, then even in the absence of faults, the ensemble cannot be synchronized with a convergence:

$$\nu < \varepsilon\left(1 - \frac{1}{N}\right) \tag{2.43}$$

where ε represents the upper bound of uncertainty in message passing latency. In cases where the message passing latency is unbounded, probabilistic clock synchronization algorithms have been found to be very useful for both external and internal clock synchronizations. For a detailed analysis of such methods the reader is directed to Cristian [3] and Landelius and Lynch [7].

NUMERICAL AND ANALYTICAL PROBLEMS

2.1. Deduce a relationship between the message passing latency and minimum value of global time granularity. Given a latency jitter of 20 μs, a clock drift rate of 10^{-5} s/s and a resynchronization interval of 1 s, calculate the precision achievable by a central master algorithm in an ensemble of 10 clocks.

2.2. A control system consists of a time-driven sensor that sends data to a controller every 10 ms over a network. The controller is also time-driven and updates data every 10 ms. The maximum latency of data transfer is 100 μs. The drift rate associated with the clocks 10^{-5} s/s and the resynchronization interval is 100 s. If the convergence of synchronization is 200 μs, by what quantum of time should the controller's timeline be skewed so that the controller in the *k*th step always picks up the data sent by the sensor during that step? Assume no data loss. What should be the smallest value of a macrotick if two consecutive samples sent by the sensor belong to two consecutive macroticks? Is the timeframe reasonable in this case?

2.3. Establish a limit for μ for an interactive convergence algorithm in terms of maximum drift rate, resynchronization period, uncertainty bound in reading clocks, and number of faulty clocks. Hint: Look up Lamport and Melliar-Smith [6].

2.4. Prove that in an ensemble of N clocks, if the maximum uncertainty of communication latency is ε, then the convergence is $\varepsilon(1 - 1/N)$ Hint: Look up Landelius and Lynch [7].

References

1. Kopetz, H. *Real-Time Systems: Design Principles for Distributed Embedded Applications*. Springer, 1997, ISBN: 0792398947.
2. LeapSecond.com. GPS, UTC, and TAI clocks. http://www.leapsecond.com/java/gpsclock.htm.
3. Cristian, F. Probabilistic clock synchronization. *Distributed Computing*, 3, 146–158, 1989.
4. Lamport, L., Shostak, R., and Pease, M. The byzantine generals problem. *ACM Transactions on Programming Languages and Systems*, 4(3), 382–401, 1982.
5. Pease, M., Shostak, R., and Lamport, L. Reaching agreement in the presence of faults. *Journal of the ACM*, 27(2), 228–234, 1980.
6. Lamport, L. and Melliar-Smith, P. M. Byzantine clock synchronization. *ACM Transactions on Programming Languages and Systems*, 4(3), 68–74, 1984.
7. Landelius, J. and Lynch, N. An upper and lower bound for clock synchronization. *Information and Control*, 62, 190–204, 1984.

3

Communication Protocols

In order that a distributed real-time system (DRTS) composed of a number of subsystems delivers a specified real-time response, the communication between the subsystems must have a finite latency. This has given rise to a number of communication protocols. In this chapter, a few of these protocols used for implementing a DRTS for various applications are reviewed.

3.1 Basic Concepts

In this section, the basic concepts and terminologies associated with communication protocols are presented.

3.1.1 Efficiency Latency and Determinism

The constituent subsystems in a DRTS communicate through messages. Sending or receipt of a message constitutes an *event*. A message has a generic format as shown in Figure 3.1.

The different fields of the generic message format represented by Figure 3.1 may be explained as follows:

SYNC	This marks the start of the message and is used for synchronization between the sender and the receiver. A typical example is the start bit in asynchronous serial communication.
HEADER	This field contains routing data, for example, source and destination addresses, priority information, and information regarding size of data being sent.
DATA	This is the actual data content of the message. It is called the payload.
ERROR CODE	Error detection code.
END	End of message. A typical example is the stop bit in asynchronous serial communication.

If the total size of the message is H bytes and the payload is D bytes, then the *bitwise efficiency* of the protocol is D/H.

Again, as seen from Figure 3.1, a message can also contain information to control the communication and such *control messages* are not useful for an application. Therefore, the *message efficiency* becomes very important and is

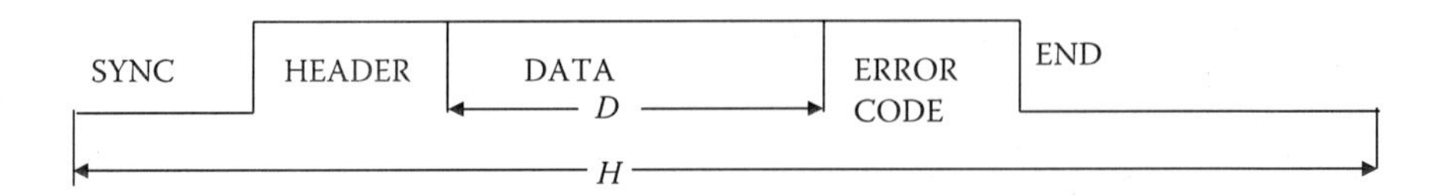

FIGURE 3.1
Generic message format.

defined as the fraction of the total number of messages useful for an application. Ideally, the bitwise efficiency and the message efficiency should both be high for an efficient communication protocol.

A protocol is usually implemented over a shared communication network and the *latency* of the message transmission is an important parameter. The *latency* is defined as the interval between the instant at which a message is queued at the source and the instant at which it is completely received (until the last bit) by the receiver. The variation in latency is known as *jitter*, as stated in Chapter 2. Another network parameter that is very important from the point of view of a DRTS is the *sustained maximum capacity* that the communication network can handle, and this is defined as the number of messages per second. Clearly, this is a function of the network *bandwidth*. The sustained maximum capacity must be more than the peak demand of the DRTS.

The next important attribute associated with a communication protocol is *media access control*, which determines how the shared network is accessed and shared between the constituent subsystems in a DRTS adhering to the protocol. Different approaches are collision sense multiple access/collision detect (CSMA/CD), collision sense multiple access/collision avoidance (CSMA/CA), and time division multiple access (TDMA). These are discussed with the individual protocols that use them.

Coupled with media access is the concept of *coding* or the manner in which the bits are sent over the network. Two basic schemes are *return to zero* (RZ) and *nonreturn to zero* (NRZ) coding. Usually a logical 1 and a logical 0 are represented by two distinct states of an electrical signal, and an RZ code forces the data to logical 0 at predefined intervals in order to synchronize between the sender and the receiver. A typical example is the stop bit, which is always a logical 1, and the following start bit, which causes a transition to a logical 0 in asynchronous serial communication. Referring to Figure 3.1 an RZ code can be achieved in the simplest case by assigning the SYNC and END bits to be complements of one another. *Manchester coding* is a typical example of RZ coding. In Manchester coding, a logical 0 is represented by a transition from 1 to 0 at the center of a bit time and a 1 is represented by a complementary transition, that is, a transition from 0 to 1. Figure 3.2 shows the signals for a sequence 0100.

A technique followed in NRZ coding to force synchronization in long sequences is known as *bit stuffing*, where a complementary bit is inserted

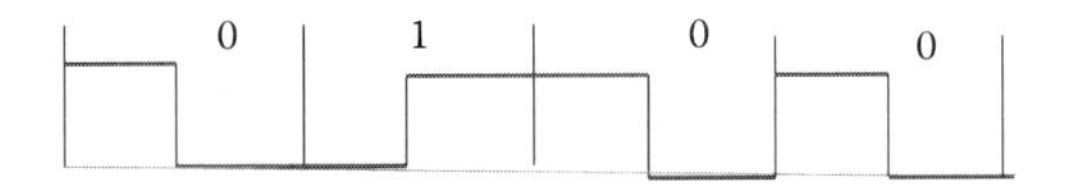

FIGURE 3.2
Manchester coding for sequence 0100.

after a sequence of finite predetermined length of a bit of a particular logic level. This, however, reduces the bitwise efficiency of the protocol.

With a bounded response time requirement, it is clear that determinism in a DRTS can only be ensured with a real-time communication protocol that ensures a bounded latency of message passing.

3.1.2 Flow Control

Flow control is an important attribute of a communication protocol for any DRTS, as it defines how the protocol detects nonreceipt of messages and what is done in case of a nonreceipt. Nonreceipt may arise because of a number of causes including buffer overflow at either the sender or the receiver. Flow control may be *explicit* or *implicit*. In explicit flow control, the sender and the receiver are each associated with a client at each end, which is the application. The sender end server sets up two counters: a *time-out* counter and a *maximum attempt* counter. The sending end server, or simply the sender, sends the message and waits for acknowledgment after starting the time-out counter. If the receiving end server, or simply the receiver, receives the data, it sends an acknowledgment to the sender only; else it does nothing. If the sender receives the acknowledgment, it informs its client that the send is complete and resets the time-out counter; otherwise it waits for the time-out value to be reached and retries send setting up the maximum attempt counter. Each time the time-out value is reached, the counter is reset until the number of retries equals the maximum attempt count. The error detection, in this case, is only done by the sending end process. On the contrary, in implicit flow control, the receiver is programmed to receive messages from the sender at some predesignated instants only. No explicit acknowledgment is sent and the error detection is only by the receiving end processes.

3.1.3 Wired and Wireless Network Topologies

The topology of interconnection in a DRTS was discussed in Chapter 2 for wired networks. However, with the advent of wireless networking, wireless topologies have also become very important, a typical example being wireless sensor networks. Logical topologies that can be implemented in a wired network can also be implemented over a wireless network. However, a wireless network has certain constraints arising out of reflection, refraction, and scattering of transmitted signals that arrive at the receiver following different trajectories from the transmitter to the receiver. Since these paths usually have different lengths, the copies arrive at different times (delay spread) and with different phase shifts at the receiver and overlap causing a significant reduction of received power, and this phenomenon is known as a *deep fade*. If the stations move relative to each other, then the number of trajectories and the resultant phase shifts vary with time. This results in a fast fluctuating signal strength at the receiver and the phenomenon is termed as *fast* or *multipath* fading. These cause a larger bit error and packet loss in a wireless network compared to a wired network. Statistical studies on packet loss and bit errors in a wireless network have been reported by contemporary researchers [1] and the results indicate the following:

- Both bit errors and packet losses occur in bursts spread over a finite interval separated by intervening error-free periods.
- The bit error rates depend on the modulation scheme. Schemes with higher bit rates are associated with higher error rates.

However, the most important phenomena that distinguish a wireless network from a wired network are *hidden* and *exposed* terminals. These are explained with Figures 3.3 and 3.4, respectively.

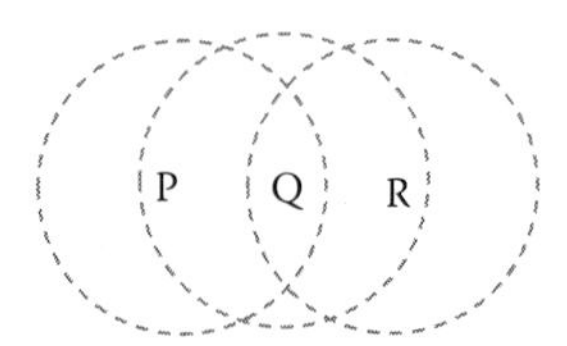

FIGURE 3.3
Hidden terminal.

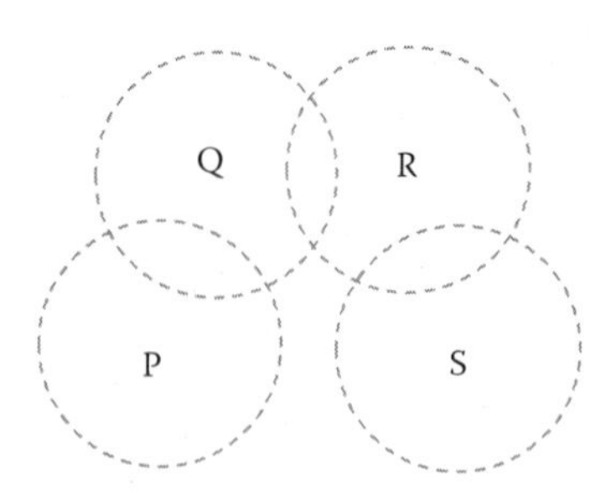

FIGURE 3.4
Exposed terminal.

As seen in Figure 3.3, the transmission radii of the three stations P, Q, and R are such that transmitters P and Q are in the range of R, but P is not in the range of R and vice versa. If R starts to transmit to Q, the transmission cannot be detected by P by its carrier-sensing mechanism and it considers the medium to be free. Consequently, if P starts a packet transmission, a collision occurs at Q. The terminals P and R remain hidden to each other. In a wired local area network (LAN) this problem does not occur.

The problem of an exposed terminal also occurs because of overlapping radii of transmitters. As shown in Figure 3.4, the transmitters P, Q, R, and S are such that only communications between the pairs {P, Q}, {Q, R}, and {R, S} are possible.

In the event Q starts a transmission to P, R cannot start a simultaneous transmission to S, as it finds the medium busy. Transmitter Q thus acts as an exposed terminal for R. Topologies for DRTS applications on a wireless network should take these factors into consideration.

3.1.4 The Protocol Tree and Application-Specific Choices

The communication protocols used for building DRTS applications follow a multilayered approach. The term *protocol* is actually a stack with multiple abstractions, as shown in Figure 3.5.

The lower-most layer represents the media access control and defines how the physical medium is handled as well as the electrical signals. The next layer defines the message formats and the flow control mechanism, and provides the interface for the various DRTS applications falling in different domains as shown in Figure 3.5.

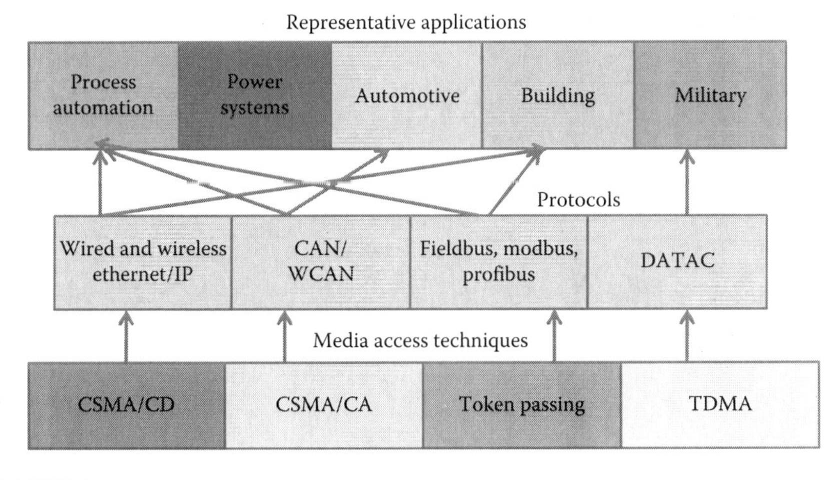

FIGURE 3.5
(See color insert.) The protocol tree.

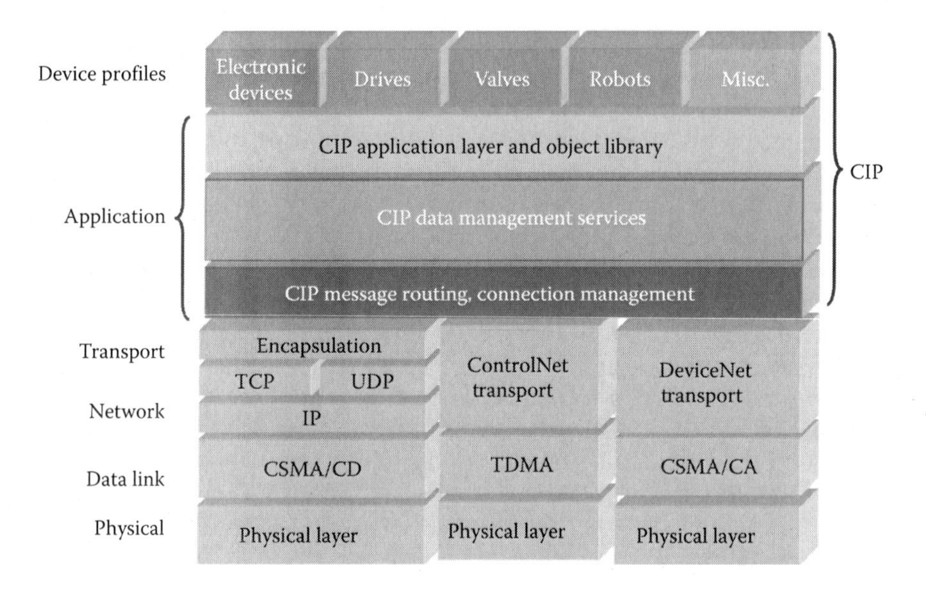

FIGURE 3.6
(**See color insert.**) The CIP in OSI context. (From "Networks built on a Common Industrial Protocol," http://www.odva.org/Portals/0/Library/Publications_Numbered/PUB00122R0_CIP_Brochure_ENGLISH.pdf.)

Often an intermediate layer is introduced between the protocol layer and the application, and this can be viewed as a *high-level abstraction,* which allows additional advantages in terms of flexibility and configurability for DRTS applications. A typical example is the Common Industrial Protocol (CIP) [2], as shown in Figure 3.6 in the Opens Systems Interconnection (OSI) context.

CIP is object oriented. CIP-based implementations use the same application layer. The application data remains the same regardless of the type of network used.

3.2 Selected Protocols

In this section, a discussion covering the different aspects of a few popular protocols are presented, covering the lowest two layers of the protocol stack represented by Figure 3.5. The protocols are discussed from three aspects: media access, efficiency, and determinism. The protocols discussed are Switched Ethernet with User Datagram Protocol/Internet Protocol (UDP/IP), Controller Area Network (CAN), Time Division Multiple Access (TDMA), Token Bus and Token Ring, FlexRay, and the Proximity-1 Space Link.

3.2.1 Switched Ethernet with User Datagram Protocol/Internet Protocol (UDP/IP)

Ethernet used for real-time communication commonly uses the 10 Mb/s standard (e.g., Modbus/TCP). Higher-speeds (100 Mb/s or even 1 Gb/s) Ethernet is mainly used in data networks. Standard Ethernet is normally avoided for real-time applications because of the nondeterministic communication delay caused by collisions when multiple stations try to access the physical medium simultaneously. In this case, a source waits for the communication line to be idle to transmit. If a collision is detected during the transmission, the source interrupts the transmission and broadcasts a jam signal to notify an occurrence of the collision to other stations. The jamming sequence is a 32-bit sequence prefixed with a preamble. Thus, a transmitter will minimally send 96 bits in the case of collision (64-bit preamble + 32-bit jamming sequence), and the frame is called a *runt* frame.

In fact, there exists a minimum size of an Ethernet frame and this is 512 bits. The existence of a minimum frame size can be understood from the following considerations: Consider a case where a node n_i starts transmitting. A node n_j located at a distance d receives the first bit of the transmission after a time of T_P s, where T_P denotes the propagation time of the electrical signal from node n_i to node n_j. Now, if it be assumed that the node n_j starts transmitting its own frame just before the first bit of node n_i reaches n_j, then the node n_i will come to know of a collision after a further T_{PR} s. Therefore, the node n_i must be transmitting for a minimum interval

$$T_{\min} \geq 2T_P \tag{3.1}$$

where T_P is the propagation delay computed using the largest permissible value of d which is 2500 m, and the corresponding value of $T_{\min}$ becomes 51.2 μs. Considering a transfer speed of 10 Mbps, this yields a minimum frame size of 512 bits.

After the collision of a frame, the source station waits for the *backoff time*

$$T_b = rnd[0,(2^{\min(k,10)} - 1)] \cdot t_s \tag{3.2}$$

based on the truncated binary exponential backoff (BEB) algorithm, and then retries the aforementioned transmission procedure. In Equation 3.2, k denotes the number of collisions in a row and for $q, r \in \mathbb{Z}^+$, $rnd[q, r]$ denotes a random integer in the interval $[q, r]$. It is clearly seen that where two nodes are waiting for a third node to finish its transmission, they will first collide with probability 1, then with probability 1/2 for $k = 1$, then with probability 1/4 for $k = 2$ and so on. This retry is continued up to 16 times, and then transmission of the corresponding frame is withdrawn.

A switch is an active device that receives the frames from any node n_i and forwards the same to its destination n_j. A switch provides a path $\{n_i, n_j\}$ between any two nodes n_i, n_j on the network. Collision occurs only if simultaneous communication is attempted along any two paths that are not disjoint, for example, along paths $\{n_i, n_k\}$ and $\{n_j, n_k\}$ simultaneously, and not otherwise. The implicit assumption is that the maximum sustained capacity of the switched network is less than the peak demand. The fact that switched Ethernet has a high message efficiency owing to the extremely small bandwidth usage for media access has motivated some researchers to consider it for real-time communication requirements. Lee and Lee [3] have presented an estimate of the maximum latency of standard and switched Ethernet and the result for switched Ethernet is presented here. As reported by Lee and Lee, the total latency for a message consisting of a frame is

$$T_m = T_{ps} + T_Q + 2T_p + T_{PR} \tag{3.3}$$

where T_{PS} and T_{PR} represent processing delays for transmission at the source and destination, respectively; and T_P is the propagation delay for the electrical signal to propagate from the source to the switch, which is proportional to the length of cable connecting the node and the switch. The factor 2 comes because of the path from the switch to the destination, which is assumed to be of the same length.

The term T_Q in Equation 3.3 represents the total frame transmission delay across the switch expressed as

$$T_Q = 2T_F + T_{IF} \tag{3.4}$$

where T_F is the frame transmission delay defined as the number of bits of the frame divided by the data rate and T_{IF} and is the interframe delay when the source waits between two frames transmitted successively. It is defined as 96-bit times in the 10BASE-T Ethernet standard.

Applications layered on Ethernet use either Transmission Control Protocol/Internet Protocol (TCP/IP) or User Datagram Protocol/Internet Protocol (UDP/IP), which constitute the fourth and fifth layers of the OSI model, respectively, as shown in Figure 3.7.

With an explicit flow control, use of TCP/IP for real-time communication is rather restricted except for some protocols, for example, the MODBUS-TCP [4]. UDP/IP offers limited service when messages are exchanged between computers, using a transfer unit called a *datagram*. The messages are divided into packets at the sending end and reassembled at the receiver end. The flow control is implicit and the service in reliable packets may be dropped. The frame-oriented nature of the protocol makes it ideal for automation applications that use small amounts of data at frequent intervals. Figure 3.8 shows an UDP/IP frame embedded in a standard Ethernet frame.

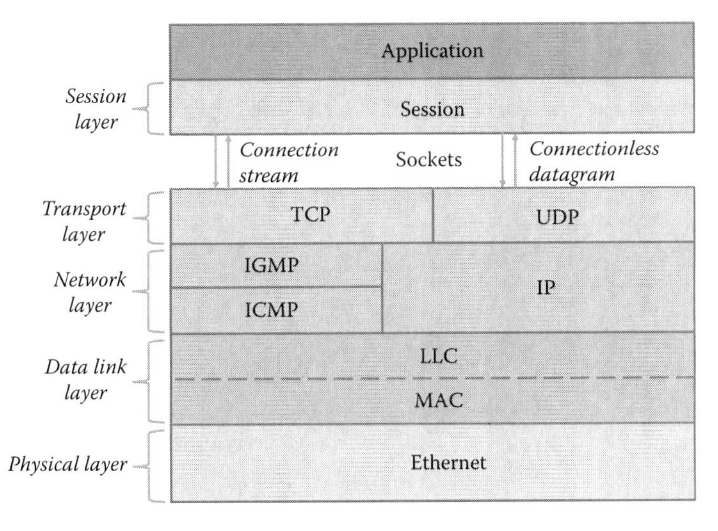

FIGURE 3.7
Ethernet over TCP/IP.

3.2.2 Controller Area Network (CAN)

The controller area network (CAN) is a deterministic protocol optimized for short messages. It was first introduced by Bosch in 1985 for replacing point-to-point interconnections in automotive control applications. The automotive industry quickly adopted CAN and, in 1993, it became the international

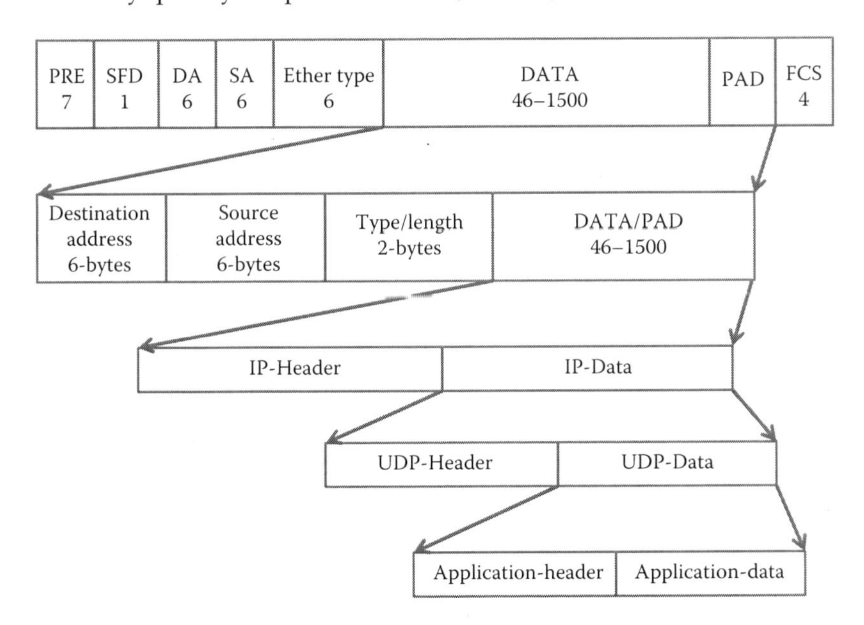

FIGURE 3.8
UDP/IP frame embedded in an Ethernet frame.

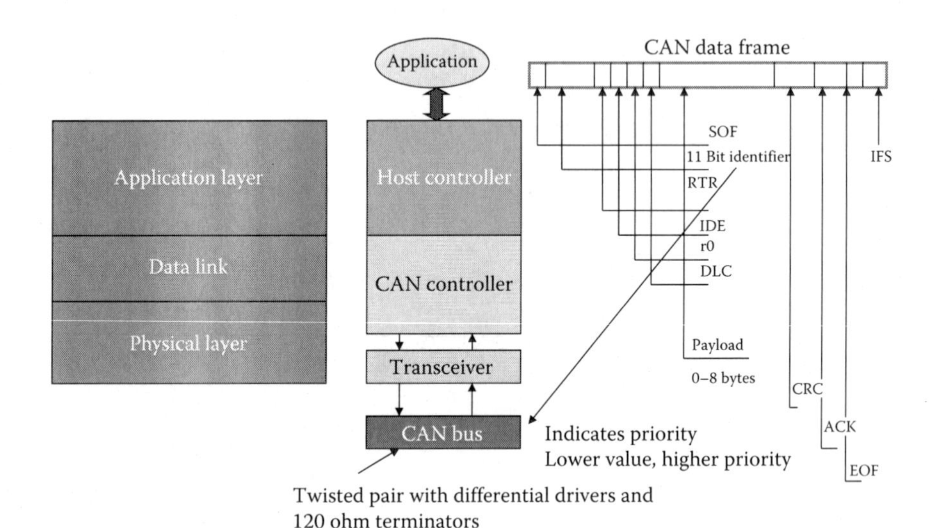

FIGURE 3.9
(**See color insert.**) Features of CAN.

standard known as ISO 11898. Since 1994, several higher-level protocols have been standardized on CAN, including CANopen [5] and DeviceNet [6].

Each device on the network has a CAN controller chip and all devices on the network see all transmitted messages. Each device can decide if a message is relevant or if it should be filtered. This structure makes a CAN flexible; additional nontransmitting nodes can be added without modification to the network.

Each message on the CAN has a priority. The message priority is specified in the arbitration field. Higher priority messages always gain access to the medium during arbitration. Therefore, the transmission delay for higher priority messages can be guaranteed. Figure 3.9 shows the relevant features of CAN.

The different fields of the CAN frame are explained as follows:

SOF (start-of-frame) bit	Indicates the beginning of a message with a dominant (logic 0) bit. It is usually a transition from a logic 1 to logic 0.
Arbitration ID	Identifies the message and indicates the priority of the message. Frames may be in two formats—standard, with 11-bit arbitration ID, and extended, with a 29-bit arbitration ID. Figure 3.9 shows a standard frame.
RTR (remote transmission request) bit	Serves to differentiate a remote frame from a data frame. A Lo RTR bit indicates a data frame and a Hi RTR bit indicates a remote frame. A remote transmission request is a message without any data, sent by one node, to request other node(s) to transmit the message of the same ID but with data.
DLC (data length code)	Indicates the number of bytes the data field contains.

Data Field	Contains 0–8 bytes of data.
CRC (cyclic redundancy check)	Contains 15-bit cyclic redundancy check code and a recessive delimiter bit. The CRC field is used for error detection.
ACK (ACKnowledgment) slot	Any CAN controller that correctly receives the message sends an ACK bit (Hi) at the end of the message. The transmitting node checks for the presence of the ACK bit on the bus and reattempts transmission if no acknowledgment is detected.
EOF	The end of frame is communicated by a 7-bit interval, consisting of a sequence of 1s. Since CAN uses bit stuffing, where a complementary bit is inserted after every six consecutive 1s or 0s in the message, this sequence does not occur in the rest of the message. The EOF, an interframe space (Hi) is also observed; the exact length of this varies depending on the length of time specified for bus idle for a given CAN controller.

Typical transmission speed is 1 Mbps up to a length of 40 m. The media access in CAN is an example of CSMA/CA and is implemented using a binary countdown mechanism. The identifier field indicates the priority of a message, which is transmitted in the binary form. As soon as a node transmits a 0, the bus is in the *dominant* state. The bus is in the *recessive* state if all the nodes transmit a 1. A node stops transmitting as soon as it tries to transmit a recessive bit when the bus is in the dominant state. The binary countdown ensures that messages with higher priority are transmitted first. This is explained with an illustrative example as follows:

Example 3.1

Two nodes start competing for the network at the same instant and try to send messages with IDs 4 and 5 (decimal) at the same instant. This is what happens:

1. Node with message ID 4 transmits a 1 (Hi) and the node with message ID 5 transmits a 1 (Hi). *The bus is now in a recessive state and both nodes transmit a 1.*
2. Node with message ID 4 transmits a 0 (Lo) and the node with message ID 5 transmits a 0 (Lo). *The bus is now in a dominant state and none of the nodes transmit 1.*
3. Node with message ID 4 transmits a 0 (Lo) and the node with message ID 5 transmits a 1 (Hi). *The bus is now in the dominant state and therefore the node with message ID 5 is stopped.*

The major disadvantage of CAN compared with the other networks is the slow data rate (typically 1 Mbs). The NRZ code also restricts the frame size. Thus CAN is suitable for small data sizes. Individual manufacturers have specific implementations of CAN and readers are referred to Watterson [7].

The transmission time of a CAN message containing D bytes of data can be approximated as [8]

$$T_D = \left\{ G + 8D + 13 + \left\lceil \frac{G + 8D - 1}{4} \right\rceil \right\} t_{bit} \tag{3.5}$$

The value of G equals 34 for a 11 bit identifier and t_{bit} is the transmission time for a single bit across the CAN. For a CAN, the worst-case response time for a message is defined as the largest time interval between the instant at which the initiating event for the message occurs and the instant at which it is received by the nodes that require it. As reported in Davis et al. [8] the response time R can be approximated as

$$R = T_j + T_w + T_D \tag{3.6}$$

where T_j represents the *queuing jitter* defined as the largest value of the time interval between the initiating event and the instant at which the message is queued, T_w is the *queuing delay* representing the maximum time a message has to wait in a queue before transmission, and T_D represents the transmission time defined by Equation 3.5. The value of T_w varies because a message may be blocked because of the following reasons:

- Lower priority messages that may be in the process of being transmitted when message is queued.
- Higher priority messages that may win arbitration and be transmitted.

Defining T_{block} as the maximum blocking time, the maximum value of the blocking time for a message can be estimated as the maximum transmission time associated with any message in the worst case, if the message is blocked by lower priority messages only. A detailed analysis leading to an estimate of the maximum queuing delay T_w is presented in Davis et al. [8].

3.2.3 Time Division Multiple Access (TDMA)

TDMA represents a media access strategy in which all nodes in a DRTS function according to a notion of global time. The channel capacity is statically divided into a number of time slots and each node is permitted to transmit during a unique slot assigned to it. The sequence of sending slots within an ensemble of nodes is called a *TDMA round*. Every node is allowed to transmit only once in every round: an empty frame is transmitted if there is no data to send and if a full frame cannot be transmitted in a slot, the transmission continues in the next scheduled slot. The TDMA nodes require fault-tolerant time synchronization with local clocks operating on a timeframe and synchronized with a predefined precision.

Header	Data 1	Data 2	Data 3	CRC

FIGURE 3.10
A TTP frame.

The Time-Triggered Protocol (TTP) defined for automotive and aerospace control applications is a typical example of TDMA. TTP follows two standards, namely, TTP/A for soft real-time requirements and TTP/C for hard real-time requirements. A typical TTP frame is shown in Figure 3.10 [9].

The application data length is variable for each node, and each data usually represents sensor data. The TTP/C frames could be one of the following types:

N-frame	These are the normal frames containing up to 240 bytes of data along with a 4 bit header and a 16 bit CRC.
I-frame	These frames are transmitted during startup and also periodically to facilitate node recovery. These frames contain up to 10 bytes of data indicating the state of the node along with a 16 bit CRC and 4 bit header.

The main disadvantage of TTP is its low protocol efficiency.

3.2.4 Token Bus and Token Ring

Token bus networks (IEEE Standard 802.4) were conceived by General Motors to meet the need for the development of a networking system to be used in its manufacturing plants—*Manufacturing Automation Protocol* (MAP). The physical topology of the network is a bus and the order in which stations are connected to the network is not important. Each station is associated with an address with the stations arranged in the form of a logical ring. The communication over this network involves use of a small frame called token.

During initialization, tokens are inserted into it in the descending order of station addresses, starting with the highest. The token is passed from higher to lower addresses. Once a station acquires the token, it has a fixed interval (slot) during which it may transmit frames. The number of frames that can be transmitted by a station during this slot will depend on the length of each frame. A frame could be a *data frame* or a *control frame*. If a station has no data to send, it simply passes the token to the next station without delay.

Token bus allows prioritization of messages by allowing four different priority levels, that is, 0, 2, 4, and 6. Each station maintains four different queues, and frames are transmitted in descending order of priority. The general format of a token bus frame is shown in Figure 3.11.

The frame control byte demarcates a frame as a data or a control frame. If the frame is a data frame, then it indicates the priority level of the message and also the flow control information, that is, explicit or implicit. The control frames are demarcated using the frame control byte and are used for initialization and maintenance of the logical ring. Table 3.1 presents the different valid combinations [10].

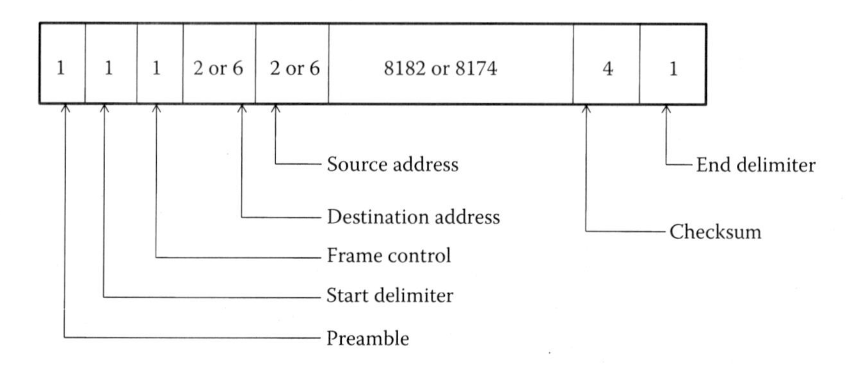

FIGURE 3.11
A token bus frame.

TABLE 3.1
Frame Control Byte in a Token Passing Network

Frame Control Byte	Service
0x00	*Claim_token*
0x01	*Solicit_successor_1*
0x03	*Who_follows*
0x0C	*Set_successor*
0x02	*Solicit_successor_2*
0x04	*Resolve_contention*
0x08	Token

The significance of the individual services is as follows:

Claim_token	When the first node on the token bus comes up, it sends a *Claim_token* packet to initialize the ring. If more than one station is active at the time of initialization, a bidding process is followed to resolve the contention.
Solicit_successor_1	This allows new nodes to join the ring. Each node in the ring periodically sends these frames with its current address and the address of its successor. Nodes whose addresses lie within this range are allowed to join the ring.
Who_follows	If a node n_i passes a token to its successor n_j it monitors the network for transmissions from n_j, which could either be a frame or a token to the successor of n_j, say n_k. If this does not happen n_i assumes that n_j has gone offline and transmits a *Who_follows* frame specifying n_j.
Set_successor	When n_k detects a *Who_follows* frame from n_i with n_j specified, it transmits a *Set_successor* frame setting itself as the successor of n_i.
Solicit_successor_2	This frame is transmitted by a node n_i if two consecutive nodes n_j and n_k go offline and there is no response, then n_i transmits a *Solicit_successor_2* frame to set its successor from among the active station following a bidding procedure.
Resolve_contention	This frame is transmitted by a node n_i whenever it needs to find a successor through a bidding process. The frame is broadcast to all active stations to initiate bidding.
Token	This frame is transmitted by a node n_i to pass the token to its successor n_j.

As bidding starts, contention may arise with multiple stations transmitting simultaneously. The contention is resolved using a binary countdown mechanism. The contending nodes send random data for 1, 2, 3, and 4 units of time depending on the first 2 bits of their address. The node that transmits for the longest time wins the bid. If contention remains unresolved, for example, in case where the first 2 bits of the address of two or more nodes are same, then the binary countdown is repeated with the next 2 bits.

Finally, if a node holding the token goes offline without passing it, then the ring initialization procedure is invoked with a *Claim_token* frame transmitted by all active stations. And if a node n_i holding a token notices a transmission from another node n_j it simply discards its token. Thus, at any point of time, there is only one token circulating in the network.

A token ring network is described by the IEEE 802.5 standard and allows up to 250 nodes to be physically connected as a ring or a star, the logical connectivity being a ring. A token ring allows two transmission speeds: 4 Mbps or 16 Mbps. Like the token bus, the token ring also uses a frame. The standard format for a token ring frame is shown in Figure 3.12.

The bits A and C in Figure 3.12 constitute the *Frame Status* field and are set to 1 if the destination receives the frame and accepts the frame, respectively.

In a token ring, a token circulates in the ring and in order to transmit; a node must possess a token. A token in a token ring is a frame consisting of 3 Bytes, as shown in Figure 3.13.

If a station wants to transmit a frame, it inverts the T bit (Figure 3.13) [10], which instantaneously changes the token into a normal data packet and thus *grabs* the token. A node can hold a token for a default maximum of 10 ms and during this interval a node sends queued frames sequentially until the holding time elapses. The node then recirculates the token. Like the token bus, a token ring allows prioritization and the priority bits define which nodes can access the token—a node can transmit only if its priority is greater than or equal to that of the token.

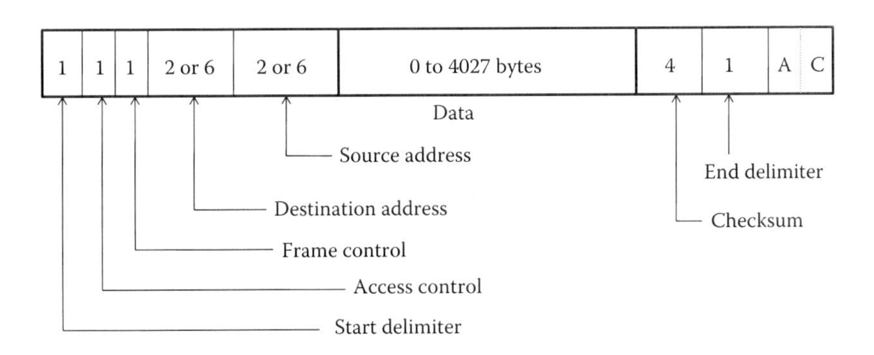

FIGURE 3.12
A token ring frame.

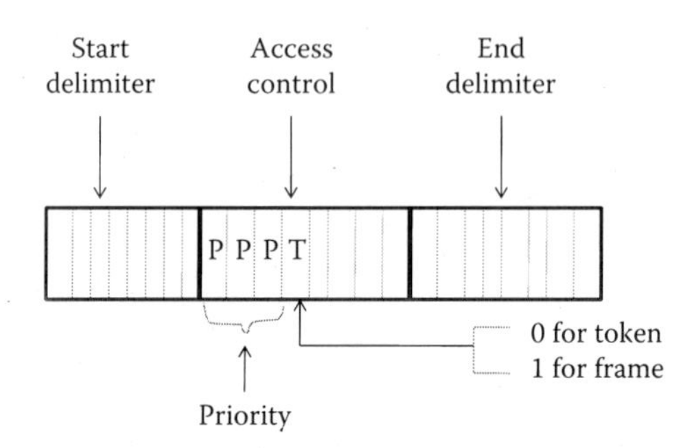

FIGURE 3.13
A token in a token ring.

In a token ring, a node can operate in any one of the following modes:

Listen mode—In this mode, a node listens to the data and transmits the data to the next node. There is a single-bit delay associated with the transmission. Details of this process can be understood by examining the IEEE 802.2 standard, which defines the Logical Link Control (LLC).

Transmit mode—In this mode, a node discards any data and puts the data on the ring.

Bypass mode—This applies to a node that is down. The node is bypassed.

Enhanced fault tolerance in a token ring network can be achieved using a *ring concentrator* with at least two twisted pair cables connecting each station to the ring concentrator: one for receiving data and one for sending. A concentrator changes the ring topology to a star, on failure of a node or breakage of the ring [10] while retaining the logical ring topology bypassing the failed or disconnected nodes. Alternatively, a hub or switch can also be used. Token bus and token ring protocols are often classified as *implicit TDMA* protocols.

3.2.5 FlexRay

FlexRay is a serial, deterministic, fault-tolerant fieldbus system for the automotive industry. The FlexRay consortium was founded in 2000 by BMW, Daimler, Motorola, and Phillips; Bosch, General Motors, and Volkswagen joined later. The consortium worked out a standard specification, which became an ISO Standard (ISO 17458-1 to 17458-5) after the consortium disbanded. The design goals were to increase the data transfer rate, the real-time capability, and resilience.

The FlexRay protocol provides flexibility and determinism by combining a *static* and *dynamic* message transmission. FlexRay allows use of both

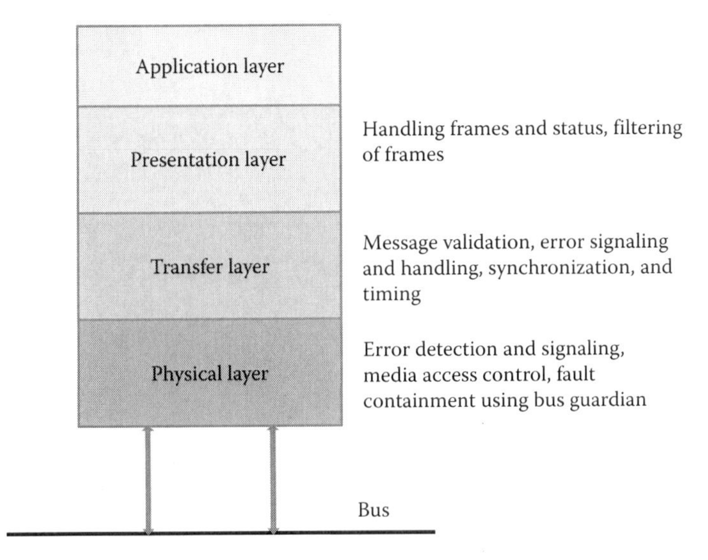

FIGURE 3.14
(**See color insert.**) The FlexRay protocol stack.

asynchronous and communication, a bounded message latency and jitter, a high data rate of 10 Mbps, redundant channels, and fault containment using node-based bus guardians. Figure 3.14 shows the layered architecture of the FlexRay protocol stack.

FlexRay supports both optical and wire communication at the physical layer level. Bus guardians help in containing faults.

In FlexRay, the flexibility comes from the options available for defining a communication cycle. Within one communication cycle FlexRay offers the choice of two media access schemes. These are a static TDMA scheme, and a dynamic minislotting based scheme. The static segment of the communication cycle is used for transmission of periodic messages like sensor data, whereas the dynamic segment is used for sending sporadic messages, diagnostic messages, and so on. FlexRay allows both event and time triggered combinations of the communication segments and three combinations are possible:

1. A communication cycle triggered by a timer event consisting of only a static segment.
2. A communication cycle triggered by a timer event consisting of a static segment followed by a dynamic segment.
3. Communication started by a start of communication (SOC) event followed only by a dynamic segment.

Figure 3.15 shows the communication cycle in FlexRay of type 2 comprising both static and dynamic segments.

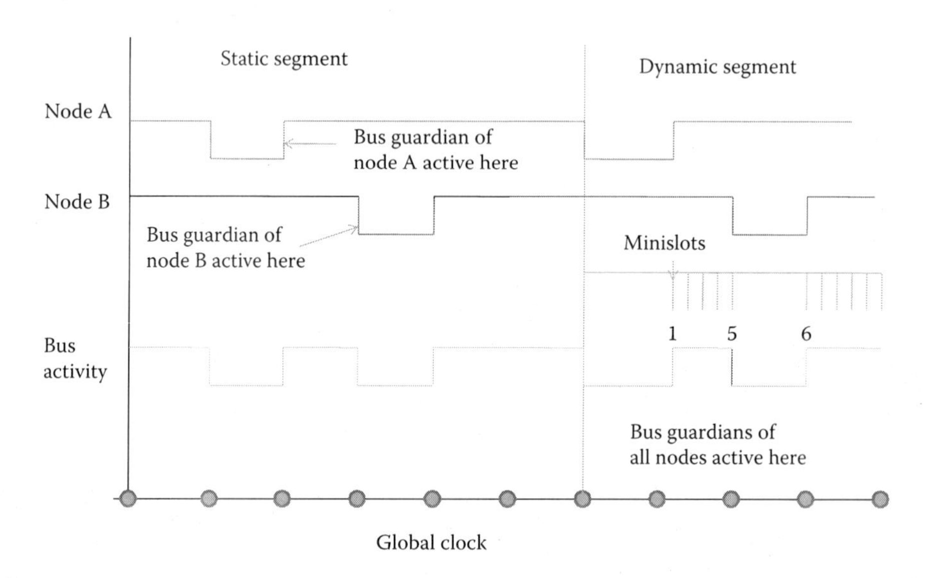

FIGURE 3.15
(**See color insert.**) FlexRay communication cycle with two nodes.

The FlexRay communication protocol is organized as a periodic sequence of communication cycles with fixed length of L slots, and the pattern is repeated after N cycles. Messages sent by nodes are associated with three attributes: (1) the cycle in which it can be transmitted C_i, (2) a periodicity P_i in terms of the number of cycles, and (3) a priority τ_i. Conflicts between messages mapped to the same dynamic segment are resolved using priorities. Messages transmitted in different cycles may be assigned the same priority and this is termed as *slot multiplexing*. Each dynamic segment in FlexRay is partitioned into equal-length slots, which are referred to as *minislots*. A slot counter counts the number of slots in the dynamic segment. At the beginning of each dynamic segment, the highest priority message gets access to the bus. It occupies the required number of minislots on the bus according to its size and the slot counter increments only by one. If the message is not ready for transmission or the size of the message does not fit into the remaining slots of the dynamic segment, then only one minislot goes empty and the slot counter is incremented by one. The bus is then given to the next highest-priority message if it is ready and the same process is repeated until the end of the dynamic segment. Further, at most, one instance of each message is allowed to be transmitted in each communication cycle. As shown in Figure 3.15, the messages sent by node A and node B are both mapped to cycle 0, and the node A message has a higher priority. An analysis of scheduling in the dynamic segment is presented by Bordoloi et al. [11].

FlexRay allows both optical and wire connectivity at a physical level. The connection topologies could be either a *passive* bus or an *active* star or a combination of both (hybrid). The faults at the transmission level are contained by the *bus guardian* module. A detailed analysis of FlexRay is available online [9].

3.2.6 Proximity-1 Space Link Protocol

Proximity-1 is a bidirectional space link layer protocol to be used by space missions, typically for *short-haul* communication between a *lander* and the *orbiter*. The lander is usually a craft like the *Sojourner* [12], which is usually controlled by the orbiter through a communication link that is short haul, compared to the *long haul* communication link between the Earth and the planetary body. The protocol has been designed by the Consultative Committee for Space Data Systems (CCSDS) and has been used extensively in all Mars missions by the National Aeronautics and Space Administration (NASA). The protocol takes into account the challenges like variable latency of the communication links, disruptive nature of the links caused by the changing radio visibility between the lander and the orbiter, power constraints, and errors in communication.

It consists of a physical layer, a coding and synchronization (C&S) sublayer, and a data link layer. This protocol has been designed to meet the requirements of space missions for efficient transfer of space data over *proximity space links*. The protocol follows a layered approach with two main layers: the *physical layer* and the *data link layer*.

The physical layer has defined functionalities both on the send side and the receive side. On the send side, the physical layer has the following functionalities:

- Provides an output bit clock to the C&S sublayer in order to receive the output bitstream
- Provides status signals to the Media Access Control (MAC) sublayer

On the send side, the physical layer provides the received bit clock/data to the coding and synchronization sublayer.

The data link layer consists of five sublayers described as follows:

1. *Coding and synchronization sublayer*—The C&S sublayer includes verification procedures and captures the value of the clock for time correlation purposes.
2. *Frame sublayer*—The frame sublayer includes frame validation procedures, such as transfer frame header checks, and supervisory data processing for supervisory frames.
3. *MAC sublayer*—The MAC sublayer defines how a session is established, and how its characteristics are modified (e.g., data rate changes) and terminated for point-to-point communications

between proximity entities. This sublayer builds upon the physical and data link layer functionality. The MAC controls the operational state of the data link and physical layers. It accepts and processes supervisory protocol data units (SPDUs) and provides the various control signals that dictate the operational state.

4. *Data services sublayer*—The data services sublayer defines the frame acceptance and reporting mechanism for proximity links on the send and the receive sides.
5. *Input/output sublayer*—The input/output (I/O) interface sublayer provides the interface between the transceiver and the on-board data system and their applications.

Figure 3.16 illustrates the layered architecture of the protocol [13].

The physical layer channel is the RF medium over which the transmission occurs. Each proximity link is associated with a *caller* and a *responder*. The link between the caller and the responder is termed the *forward link*, and the link between the responder and the caller is termed the *return link*. A caller imitates a communication link between itself and the responder on a prearranged communication channel, and the persistent activity to establish the link is called *hailing*. The details of link establishment, control, and

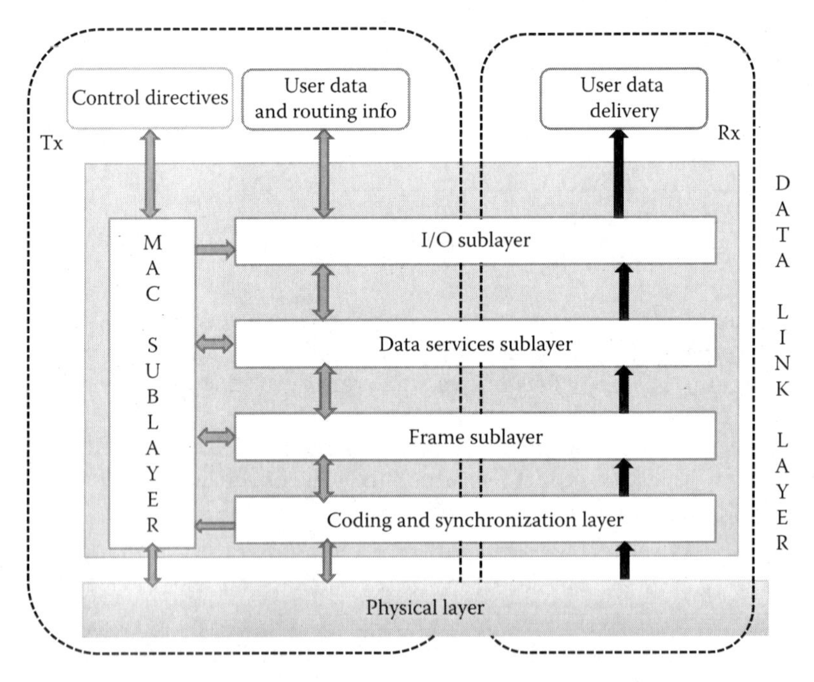

FIGURE 3.16
(**See color insert.**) Layered structure of the Proximity-1 protocol.

transmission are detailed by the Consultative Committee for Space Data Systems [13].

3.3 Multihomed and Multistreaming Approaches

A device that has multiple connections to the network is said to be *multihomed*. It can be reached via multiple addresses. A smart phone typically is a multihomed host on the Internet, since it can connect to the Internet via WLAN or via the cellular networks used by mobile phones (GSM). Servers may have multiple network interfaces connected to different Internet service providers. Multihomed hosts can be connected via multiple independent paths. This increases reliability. If one path is down or congested, the host can switch to another network interface.

Failover is only one application of multihoming. Applications that are hungry for bandwidth can use multiple interfaces simultaneously to exploit the aggregate bandwidth of all available paths.

Multistreaming is the capability of transmitting multiple streams of chunks in parallel. For instance, the text and the images on a web page might be separate streams that can be processed independently in parallel by the receiver.

It depends on the protocols to what extend the features described here can be exploited in practice. The Internet protocol stack typically uses TCP as the transport layer protocol. TCP provides reliable data transfer and strict order-of-transmission data delivery. It is a byte-stream oriented protocol, that is, the data transfer is a single stream of bytes. TCP is connection-oriented. A connection has exactly two endpoints, referred to as sockets. An endpoint is described by an IP address and a port number. Obviously, TCP does not support multihoming.

Head-of-line blocking is a consequence of the single-byte-stream nature of TCP. Figure 3.17 illustrates the problem. It shows a switch with four inputs

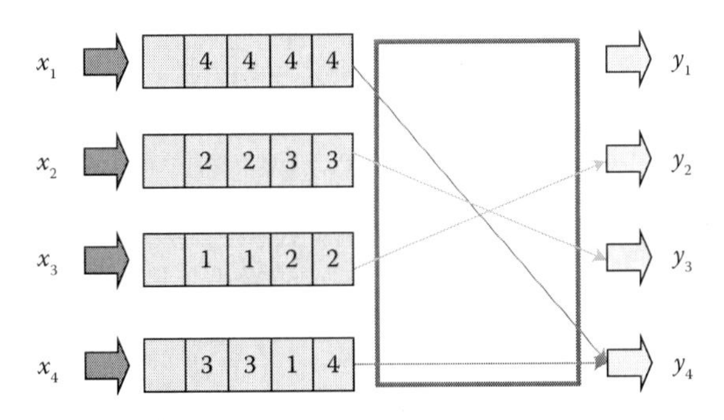

FIGURE 3.17
(See color insert.) Head-of-line blocking in TCP.

(x_1, ..., x_4) and four outputs (y_1, ..., y_4). For the elements in the input queues the output numbers are displayed. In x_1 a series of bytes waits to be output to y_4. Assume the switch has established the connection $x_1 \rightarrow y_4$. Then the head of queue x_4 will remain blocked because y_4 is busy. As a consequence, subsequent elements of x_1 are also blocked although their output ports are free.

The Stream Control Transmission Protocol (SCTP) overcomes some of the restrictions of TCP. SCTP is an Internet standard (RFC 4960). It is a message-based protocol that supports multistreaming. Data can be delivered as chunks in independent streams in order to avoid head-of-line blocking. Endpoints can have multiple addresses, that is, multihoming is possible. Multihoming is used to increase reliability. Reachability of the addresses is checked by a heartbeat mechanism. For each endpoint, there is a primary and a redundant address. If the heartbeat mechanism indicates a problem with the primary address, a failover to the redundant address takes place. Concurrent multipath communication is not supported by SCTP itself but has been implemented in a research project [14]. Multihoming and multistreaming approaches can be used to implement enhanced communication reliability in DRTS applications.

3.4 Protocol Analysis with a Protocol Analyzer

In a DRTS application, it is often necessary to analyze data packets sent over a network. This is achieved using a *protocol analyzer,* also known as a *network analyzer.* This could be either a software module or a hardware module that intercepts the packets passing through a data network and generates a time-stamped trace of these packets. A protocol analyzer usually comes with a bundled utility that allows the packet contents to be analyzed. Software modules used for analysis, like *Wireshark* [15], are usually housed on a node and add to the latencies seen at the application layer. On the other hand, hardware protocol analyzers act in a *nonintrusive* manner and do not add to the latency. Usually, they may be connected to a port of a network switch to monitor traffic across the switch (or on the network itself as a node) or between the switch and a node for monitoring transmissions to and from the node.

Figure 3.18 shows the log captured by a JDSU 6803A distributed network analyzer [16] for monitoring network traffic across an Ethernet network with UDP/IP.

Figure 3.18 represents the log of network traffic across a 100 Mbps Ethernet link between two stations with IP addresses 172.16.1.122 and 172.16.1.123. The *Time* field reports the time-stamp of the instant at which the packet is detected by the switch with a high-resolution timer. The data frame can be analyzed by selecting the appropriate field, for example, the Ethernet header as shown in Figure 3.18.

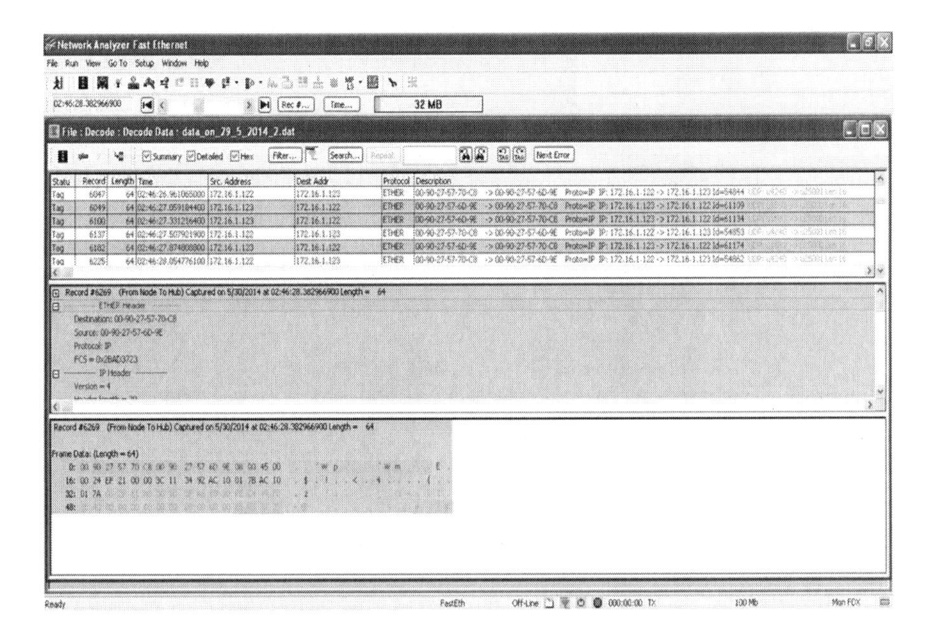

FIGURE 3.18
(**See color insert.**) Representative log from a protocol analyzer.

NUMERICAL AND ANALYTICAL PROBLEMS

3.1. Calculate the minimum and maximum bitwise efficiency of a token B = bus protocol. What is the theoretical maximum reduction in bitwise efficiency for a 256 Byte message if bit stuffing is used?

3.2. A token ring consists of six stations that communicate sequentially one after another. Calculate the maximum message efficiency. How does this compare with a scenario where the nodes communicate over switched Ethernet?

3.3. Sketch the Manchester coded signal for the bit sequence 10101100.

3.4. Two Ethernet nodes are separated by a distance of 10 m. If the minimum size of a frame is 512 bits, what is the maximum permissible transmission speed possible?

3.5. A CAN bus uses an 11 bit identifier and has a speed of 1 Mbps. Calculate the theoretical maximum contention resolution time for bus access. Under what scenario does this occur?

3.6. Calculate the message latency for a CAN message with a payload of 8 Bytes if the protocol uses a 11 Byte identifier and a transmission speed of 1 Mbps.

References

1. Cavers, J. K. *Mobile Channel Characteristics*. Kluwer Academic Publishers, 2000.
2. ODVA. Networks built on a Common Industrial Protocol. http://www.odva.org/Portals/0/Library/Publications_Numbered/PUB00122R0_CIP_Brochure_ENGLISH.pdf.
3. Lee, K. C. and Lee, S. Performance evaluation of switched Ethernet for real-time industrial communications. *Computer Standards and Interfaces*, 24(5), 411–423, 2002.
4. Simply Modbus. Modbus TCP/IP. http://www.simplymodbus.ca/TCP.htm.
5. CANopen Solutions.com. CANopen Basics—Introduction. http://www.canopensolutions.com/english/about_canopen/about_canopen.shtml.
6. RTA. DeviceNet. http://www.rtaautomation.com/technologies/devicenet/.
7. Watterson, C. Controller Area Network (CAN) implementation guide. http://www.analog.com/media/en/technical-documentation/application-notes/AN-1123.pdf.
8. Davis, R. I., Burns, A., Bril, R. J., and Lukkien, J. J. Controller Area Network(CAN) schedulability analysis: Refuted, revisited and revised. *Real-Time Systems*, 35(3), 239–272, 2007.
9. Hansen, F. O. Introduction to TTP and FlexRay real-time protocols. http://staff.iha.dk/foh/Foredrag/TTPFlexRay-LunaForedrag.pdf.
10. Sanghi, D. Computer Networks (CS425). http://www.cse.iitk.ac.in/users/dheeraj/cs425/lec07.html.
11. Bordoloi, U. D., Tanasa, B., Eles, P., and Zebo, P. On the timing analysis of the dynamic segment of the FlexRay. *Proceedings of the 7th IEEE International Symposium on Industrial Embedded Systems*, 94–101, June 2012.
12. Gerard, S. Data communication and Mars missions. http://www.marssociety-europa.eu/wp-content/uploads/2012/01/EMC12-StephanGerard1.pdf.
13. Consultative Committee for Space Data Systems. Proximity-1 Space Link Protocol—Rationale, Architecture, and Scenarios. 2013. public.ccsds.org/publications/archive/210x0g2.pdf.
14. Hassani, R., Malekpour, A., Fazely, P., and Luksch, P. High performance concurrent multi-path communication for MPI. *Lecture Notes in Computer Science*, 7490, 285–286, 2012.
15. Wireshark. About Wireshark. https://www.wireshark.org/about.html.
16. JDSU Distributed Network Analyzer. http://www.viavisolutions.com/sites/default/files/technical-library- -files/dnaoverview_ds_nsd_tm_ae_0.pdf.

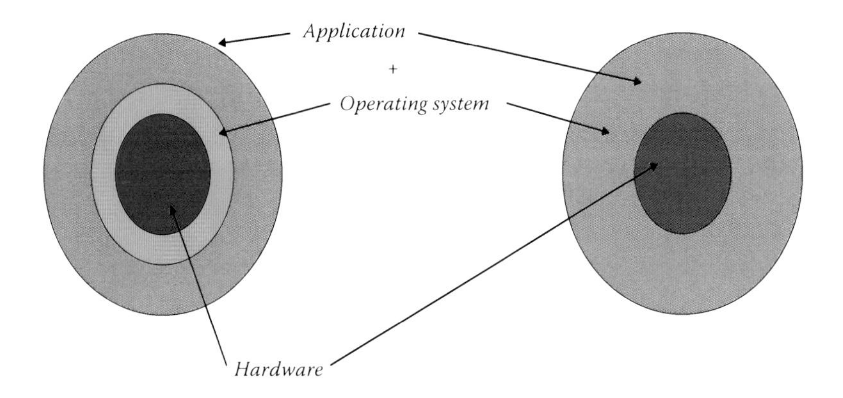

FIGURE 1.1
Difference between an embedded and a nonembedded system.

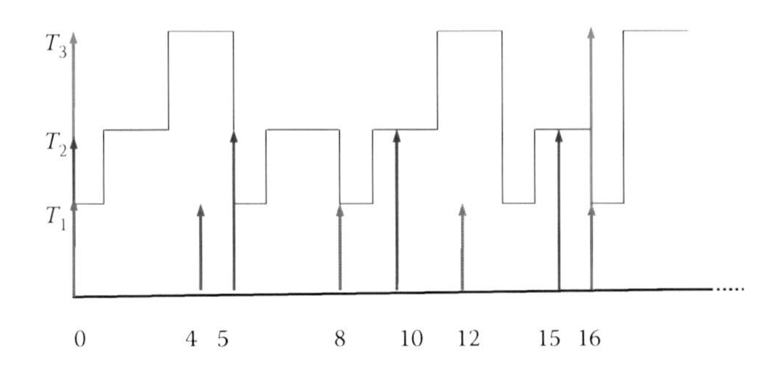

FIGURE 1.7
A representative execution profile with EDF.

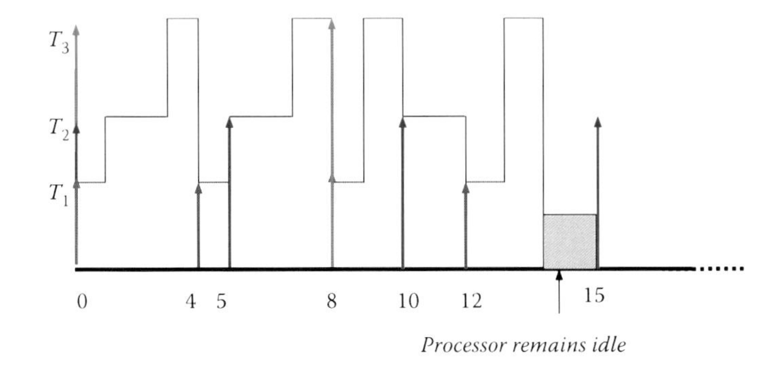

FIGURE 1.8
Execution profile with RM scheduling corresponding to Figure 1.7.

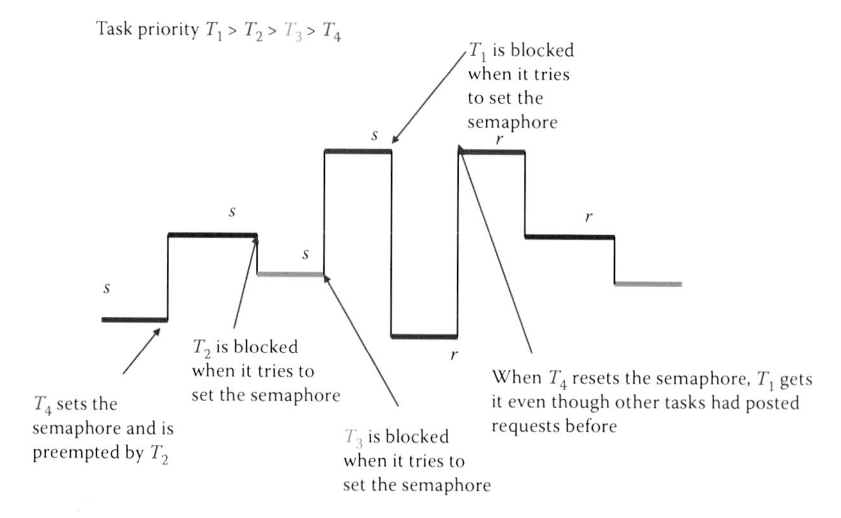

FIGURE 1.15
Burst mode semaphore.

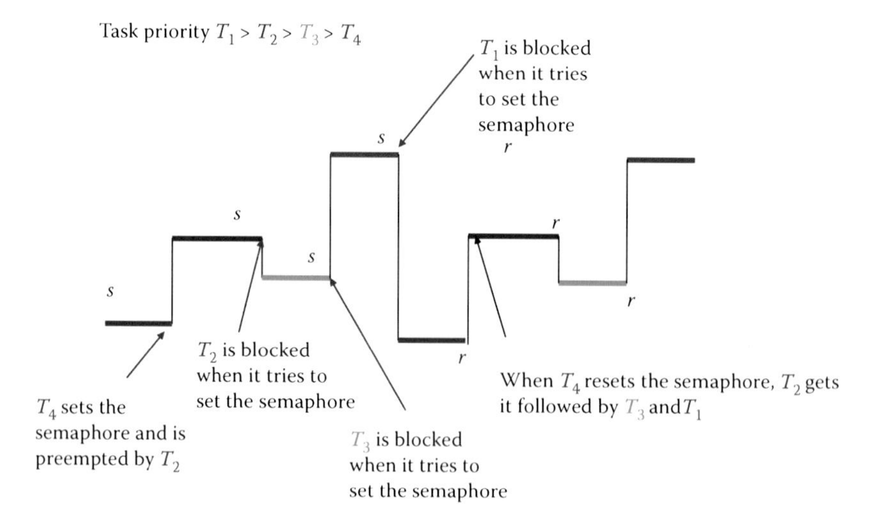

FIGURE 1.16
FIFO mode semaphore.

Task priority $T_1 > T_2 > T_3$

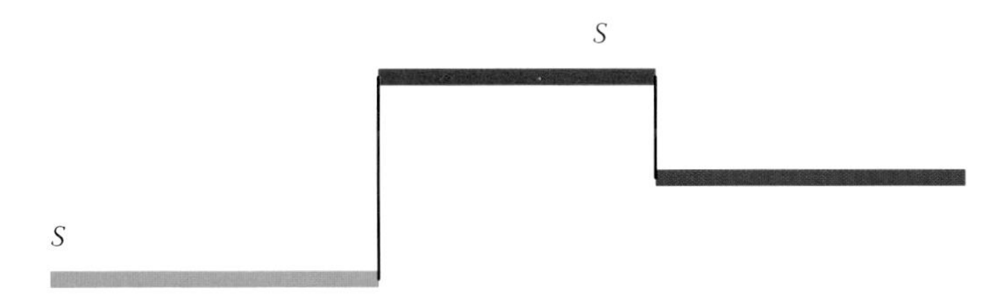

FIGURE 1.17
Execution profile under priority inversion.

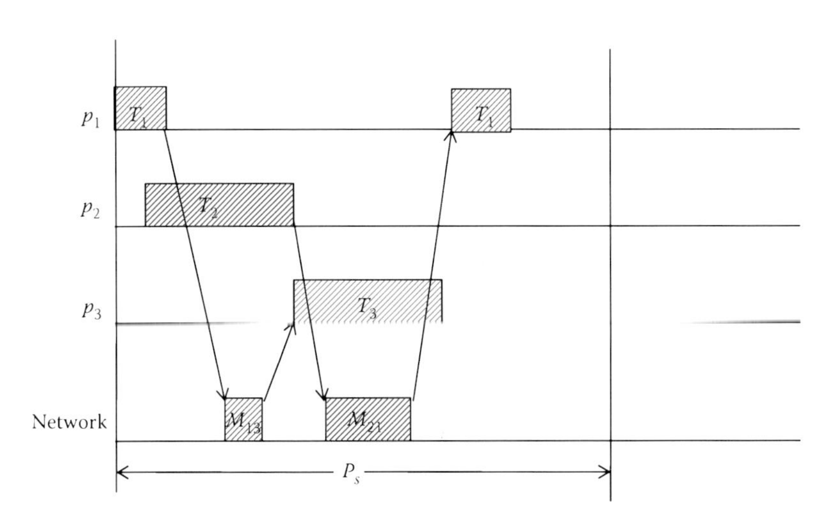

FIGURE 1.18
Coscheduling in DRTS.

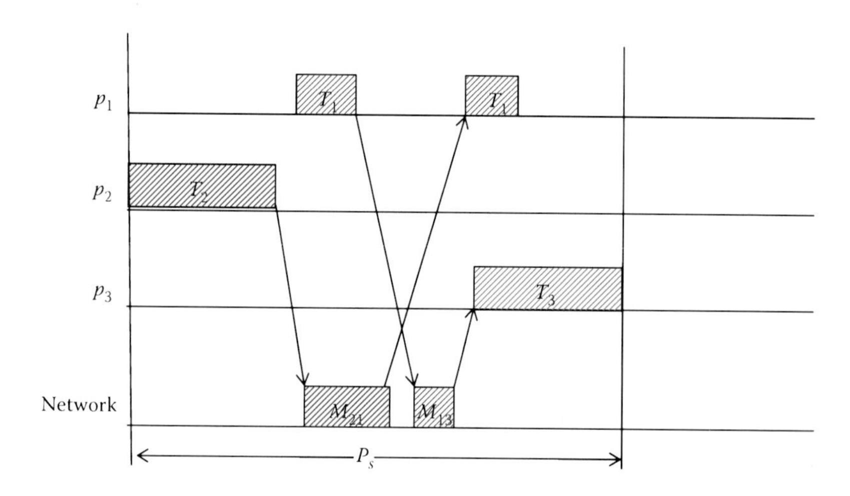

FIGURE 1.19
An alternate schedule.

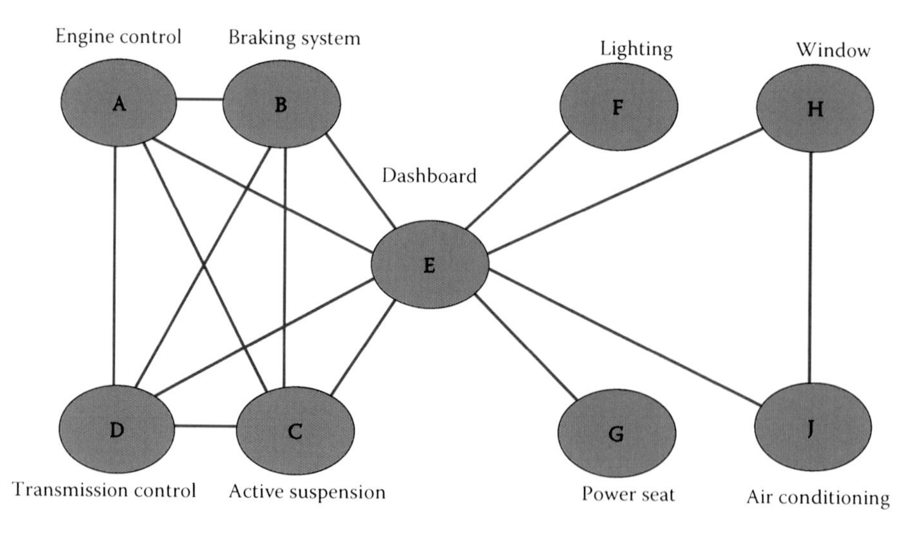

FIGURE 2.2
Automotive control-centralized scheme.

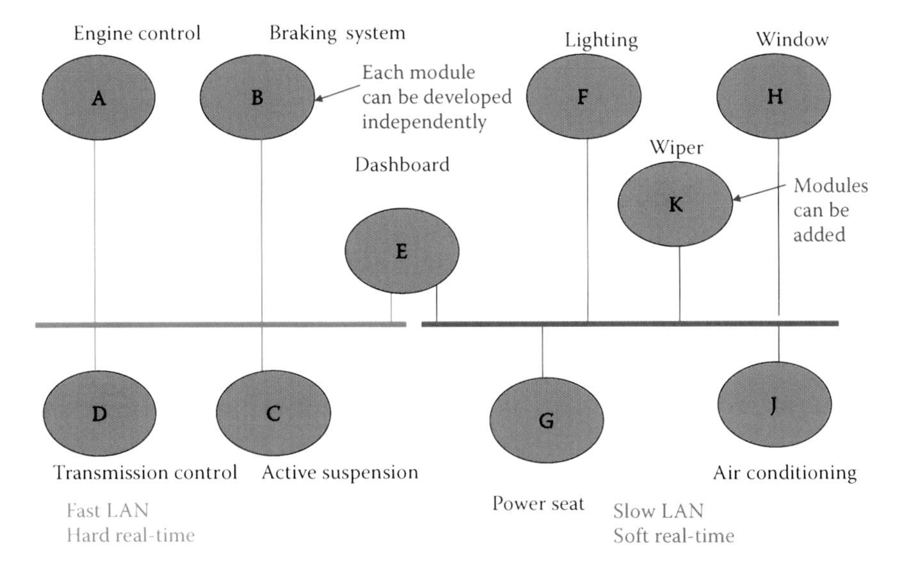

FIGURE 2.3
Modularity and flexibility in a DRTS.

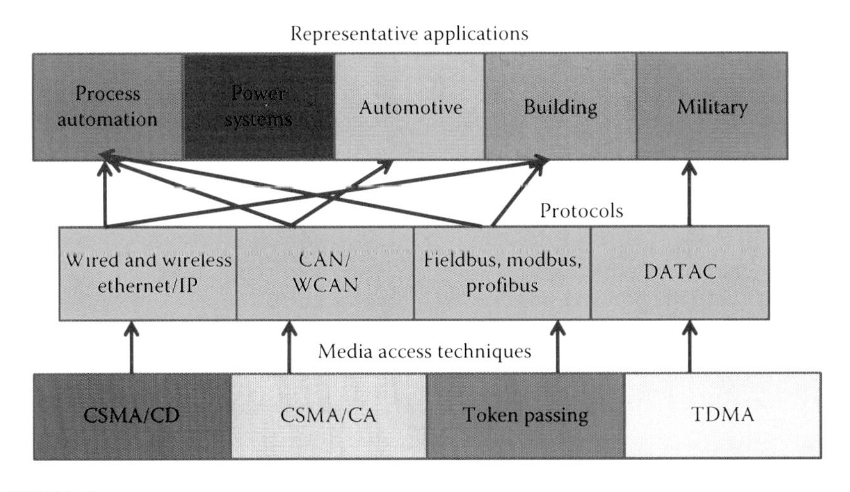

FIGURE 3.5
The protocol tree.

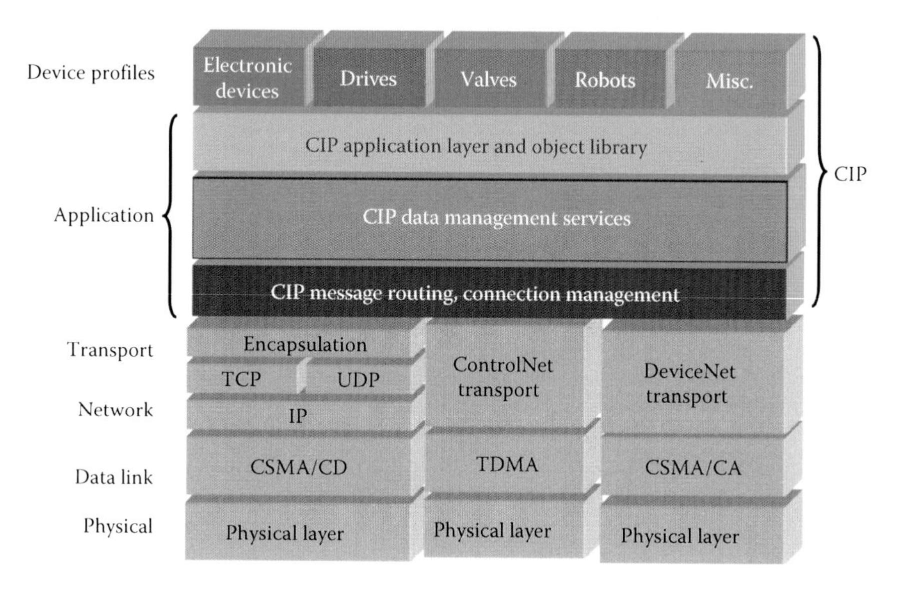

FIGURE 3.6
The CIP in OSI context. (From "Networks built on a Common Industrial Protocol," http://www.odva.org/Portals/0/Library/Publications_Numbered/PUB00122R0_CIP_Brochure_ENGLISH.pdf.)

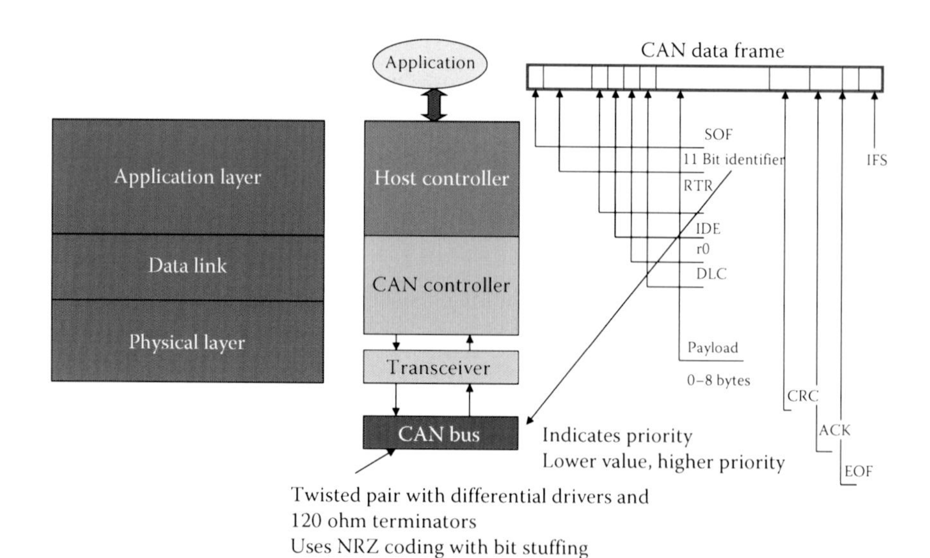

FIGURE 3.9
Features of CAN.

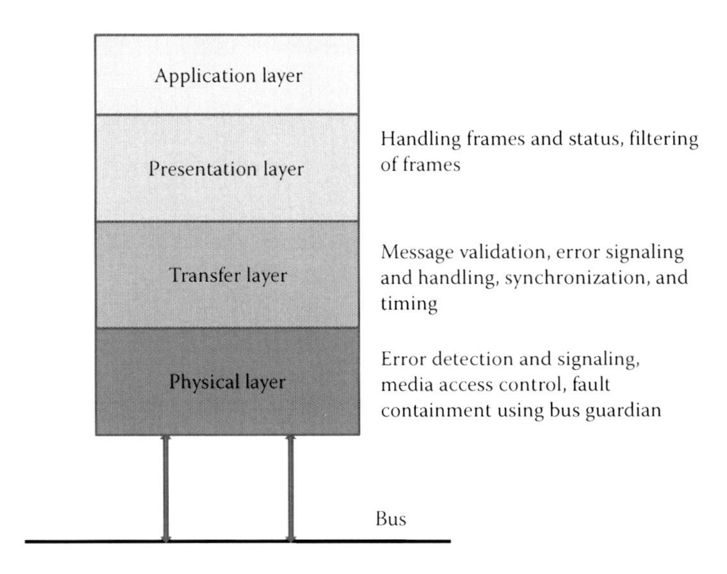

FIGURE 3.14
The FlexRay protocol stack.

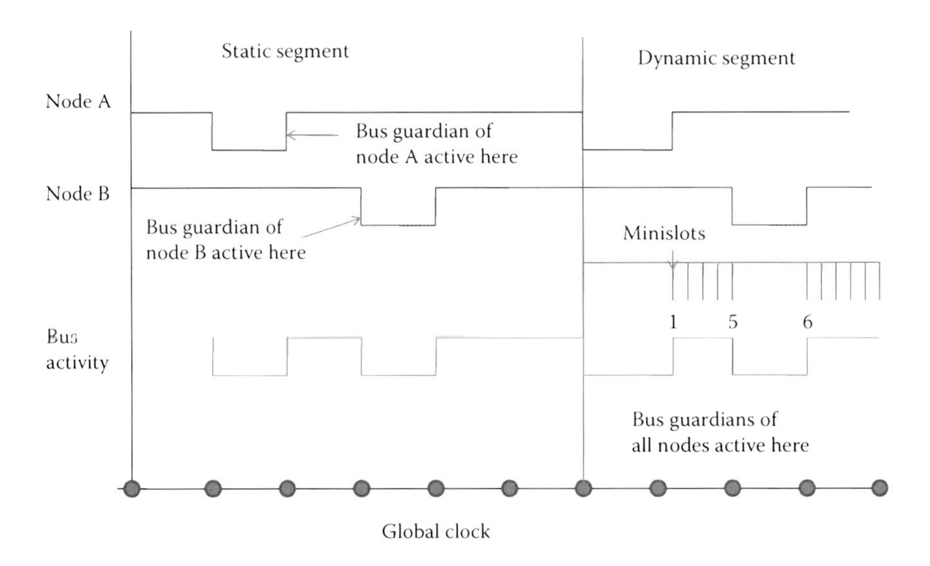

FIGURE 3.15
FlexRay communication cycle with two nodes.

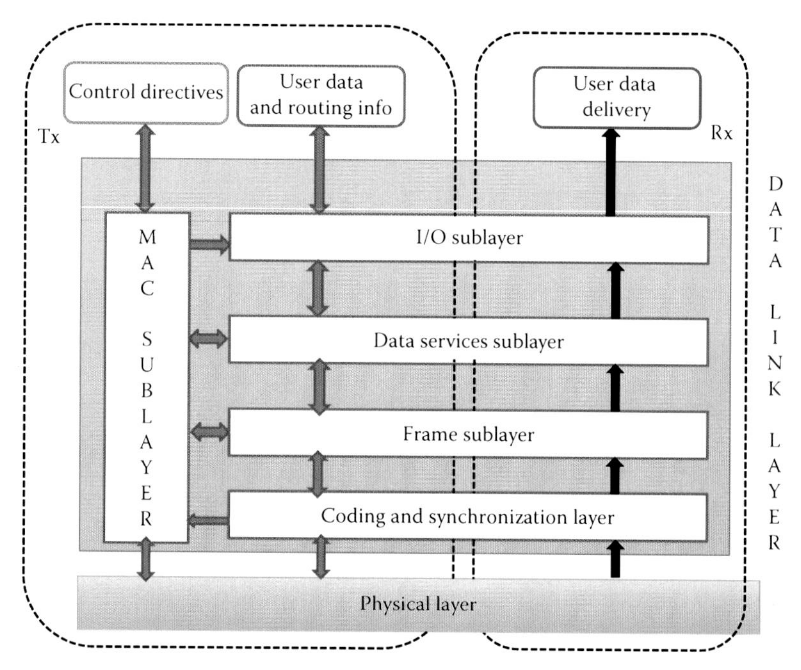

FIGURE 3.16
Layered structure of the Proximity-1 protocol.

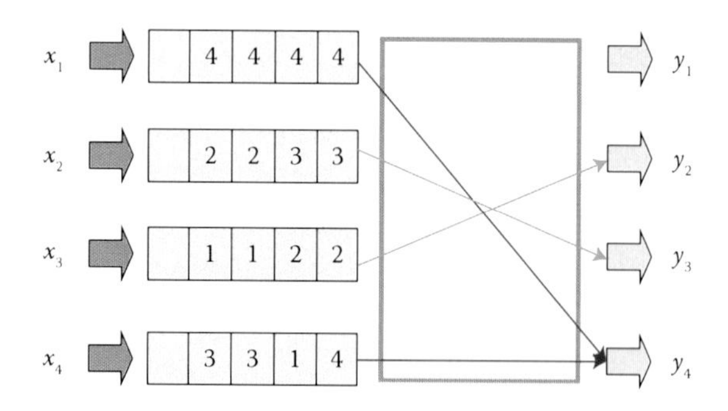

FIGURE 3.17
Head-of-line blocking in TCP.

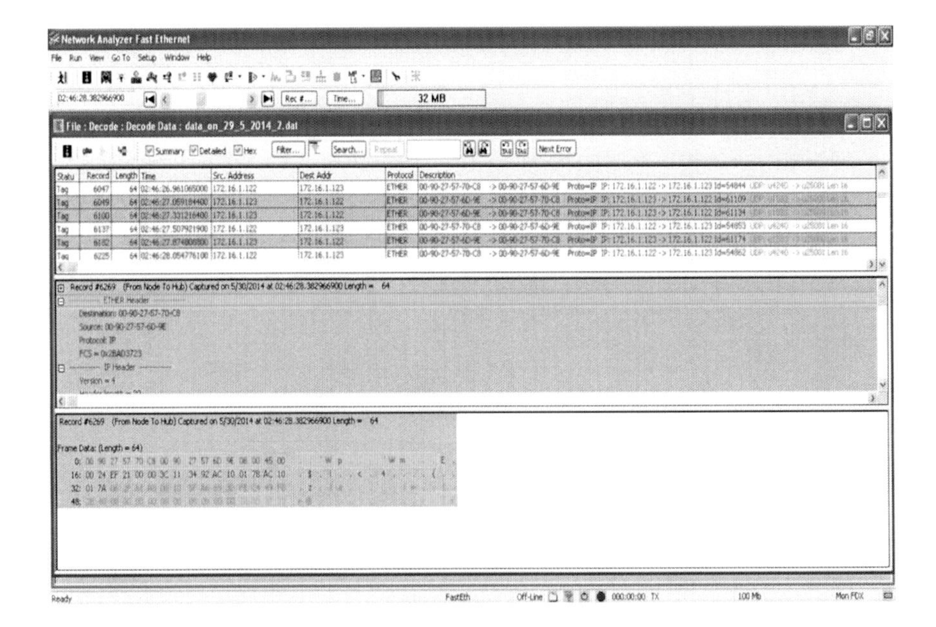

FIGURE 3.18
Representative log from a protocol analyzer.

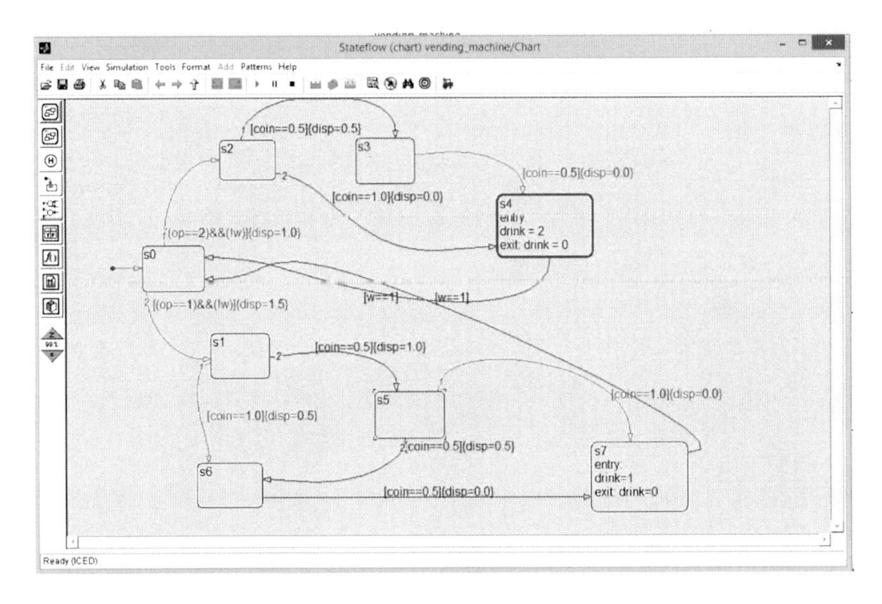

FIGURE 4.9
State activation in the chart of Figure 4.7 corresponding to the model in Figure 4.8.

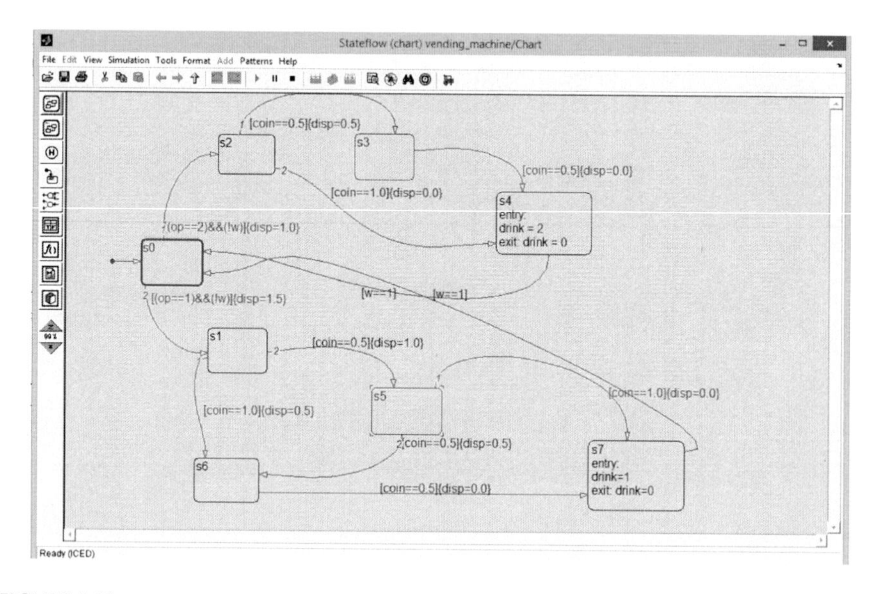

FIGURE 4.10
State activation in the chart of Figure 4.7 corresponding to the model in Figure 4.8 during retraction of bottle.

(a)

FIGURE 4.11
(a) Track with crossroad.

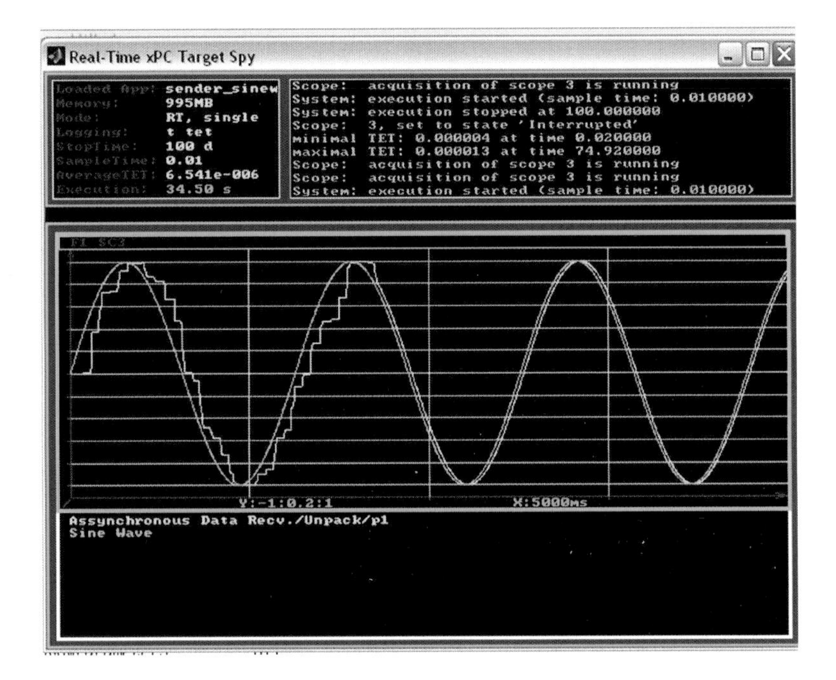

FIGURE 5.11
The transmitted and received sine waves on the Send-Receive module.

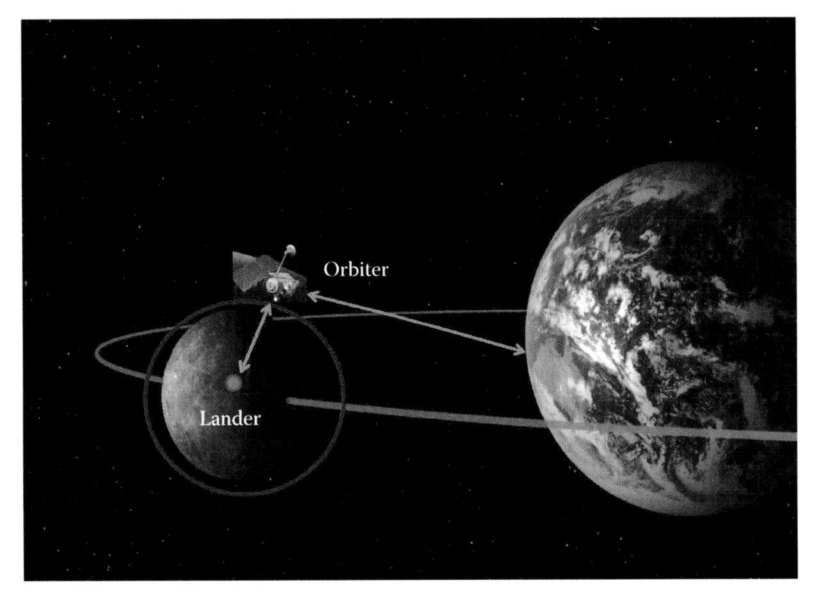

FIGURE 5.12
Communication between earth station, orbiter, and rover.

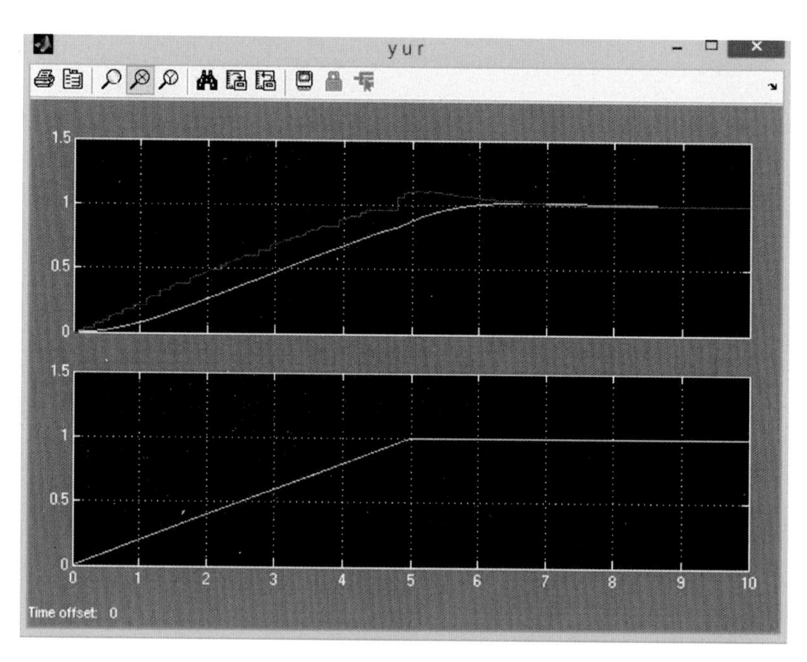

FIGURE 5.19
Output of the NCS represented by Figure 5.18.

4

Designing Real-Time and Distributed Real-Time Systems

Like the design of any type of system, the design of a distributed real-time system is a creative process with a wide spectrum of options. If a system is designed from scratch, there is a wide design space that is bound by constraints. Constraints reduce the number of design alternatives to be considered. Unfortunately, not all constraints are known at the beginning of the design process. In practice, computer control is a feature that is often added to a system that, up to now, has been run without the help of computers. An example is a production line that is modernized. The basic architecture of that plant is likely to be retained. Thus, the design of the real-time control is faced with additional constraints, which are the result of the design process of the original production line.

Kopetz distinguishes two design styles: model-based design and component-based design [1]. Model-based design takes a top-down approach. A mathematical model of the dynamics of the system (e.g., a production line) to be controlled is derived, which is used to synthesize the control algorithms. Executable models of the production line and the control algorithms are integrated into a simulation model that allows validating the system. The simulation will typically run in simulated time, which is referred to as *software-in-the-loop*. In *hardware-in-the-loop*, the subsystems of the simulation are the real target hardware and the simulation must be executed in real-time.

Component-based design takes a bottom-up approach. It starts with a collection of prefabricated building blocks. These components have well-defined and validated interfaces and properties (e.g., function and timing behavior). These components can be composed to form more complex building blocks. Their properties are determined by the properties of the components. In practice, model-based and component-based designs are combined in sort of a meet-in-the middle approach. The successive refinement of model-based design stops if the objects can be matched with predefined components. MATLAB supports component-based design by providing a large number of library components for various application domains. MATLAB components can be library routines executed on the CPU (software-in-the-loop) or software that runs, for example, on the xPC Target, an x86-based real-time system (hardware-in-the-loop).

4.1 Time- and Event-Triggered Systems

A *trigger* is an event that causes the start of some action in the control system. The action may be the execution of a task reading a variable and computing a new value of a correcting variable, or the sending of a message reporting current values of variables like pressure or temperature.

In *event-triggered* control, an action is started only if a significant event occurs. For instance, a sensor would send a message only if the temperature has changed by more than 3°C since the last message was sent. In *time-triggered* control, all actions are initiated periodically by a real-time clock. The sensor from our example would send a message every clock cycle even if the temperature remains constant.

Time-triggered control generates more communication and processing than event-triggered control. In a distributed system with event-triggered control, a component that does not receive a message from some other component does not know whether there has not been any significant event for a long time or whether the other component has failed or has become disconnected. The extra messages sent with time-triggered control act as "heartbeats," that is, they tell the receiver that the sending component and the connection to that component are still alive.

Distributed time-triggered real-time systems typically synchronize the clocks of all nodes to form a global clock. This establishes a global time base. Events can be time-stamped with the clock tick of their node. The time stamp allows defining an order between observations made on arbitrary nodes. The granularity of the global clock must be fine enough so that subsequent events get different time stamps.

4.2 Task Decomposition

When decomposing a system into components, components should be designed to perform a self-contained function that has a clearly defined external interface. A well-designed component has high internal coherence and low external interface complexity. Both the function and the interface must be clearly defined, including the allowed range of operation and timing characteristics.

4.2.1 Data and Control Flow

The decomposition of the system determines the control and data flow between components. In a distributed system, control information and data is exchanged between components by sending messages. The component interfaces set the timing requirements for communication and synchronization.

4.2.2 Cohesion Criteria and Task Structuring

Components should have strong internal cohesion, that is, each component should perform a set of well-defined, self-contained functions that form a logical unit. The component can be implemented as a single-threaded process that executes the tasks sequentially or as a multithreaded process, where threads execute concurrently. The latter is to be preferred if the tasks to be performed are independent and concurrent in nature. Possible delays in one task do not impact execution of other tasks, which would be the case with single-threaded execution at some predefined ordering.

4.3 Finite State Machines: Timed and Hybrid Automata

4.3.1 Finite State Machines

A finite state machine (FSM) can be used as a formal representation of an even-driven (reactive) system. An FSM has a finite set of states $\mathbf{S} = [s_0, s_1, \ldots, s_N]$, an input alphabet **I**, an output alphabet **O**, and a state transition function $d{:}\mathbf{S} \times \mathbf{I} \rightarrow \mathbf{S}$. At any time, the FSM is in exactly one state, s_i. Initially, the FSM is in the start state, s_0. A subset of the states can be defined as *accepting states*. The transition function, d, determines the next state s_j for a given state $s_i \in \mathbf{S}$ and a given input $k \in \mathbf{I}$. The notation $d(s_i,k) = s_j$ means if the FSM is in state s_i and reads input k, then it will move to state s_j.

A state machine is *deterministic*, if the transition function specifies exactly one next state for every pair of current state and input. If there are pairs of current state and input for which the transition function specifies two or more next states or no next state at all, the state machine is nondeterministic.* A nondeterministic FSM can be transformed into an equivalent deterministic FSM.

Depending on the purpose of their use, different classes of FSMs are distinguished: acceptors and recognizers, and transducers.

4.3.2 Acceptors and Recognizers

An acceptor checks whether a sequence of input symbols is an element of the language that the state machine accepts. If it is, the state machine reaches an accepting state, otherwise it ends up in the error state. Acceptors are used in frontends of compilers as *lexers* or *parsers*. As an example, Figure 4.1 shows

* Representations of deterministic state machines usually do not show transitions for all possible inputs. They have an implicit error state. All inputs not listed for a current state result in a transition to the error state. Subsequent input leads to a transition to the error state. The error state and the transitions to it are usually not shown. Such a state machine is deterministic, although it seems that for some input symbols there is no transition.

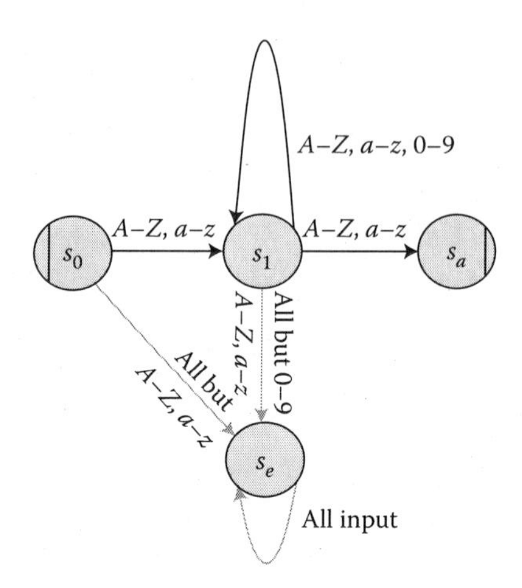

FIGURE 4.1
Acceptor.

the graphical representation of a state machine that accepts identifiers of an imaginary programming language.

Legal identifiers start with a letter, followed by an arbitrary number of letters or digits, and end with a letter. For instance a0f, v8g653q, and ab are legal identifiers, while a, 0bs, and x0 are not. The start state of the FSM is s_0, the accepting state is s_a. If in s_0 a digit is read, the state-machine goes into the error state s_e, since an illegal identifier has been detected. Note that the example FSM is nondeterministic. The error state and the transition to it often are omitted in graphical representations. Acceptors often do not generate any output.

4.3.2.1 Transducers

A transducer reads input and generates output. Transducers are used in control applications and in computer linguistics. In control applications, a transition is considered as a set of action that is executed when a condition is met or when an event is received. In some representations of FSMs, actions are associated with a state: there is an entry action that is performed when entering the state, and there is an exit action that is performed when leaving the state.

In control applications, two types of transducers are distinguished. In *Moore machines*, the output depends only on the state, that is, there are only entry actions. The output function **z** has the state as its only parameter, that is, $\mathbf{z}:\mathbf{S} \rightarrow \mathbf{O}$. In *Mealy machines*, the output depends on the input and the state, that is, $\mathbf{z}:\mathbf{S} \times \mathbf{I} \rightarrow \mathbf{O}$.

As an example of a transducer, consider a vending machine that sells water and cola in cans. Water costs 100 € per can, Coke 150 €. Assume the vending machine accepts only 50ct and 1€ coins. The machine is started by selecting one of two buttons labeled "water" and "coke." The machine then displays the price and accepts coins until the price is paid. It does not return chance. As soon as the amount of money inserted is equal to or greater than the price of the selected item, the can is sent to the collection tray. If the can is taken out of the tray, the machine becomes ready for the next purchase. The FSM is presented in Figure 4.2.

The inputs are "coke" or "water" (button pressed), insertion of coins (50ct and 1€), and "can taken out" (signal provided by a sensor in the collection tray). The outputs are the display of the price and the can in the collection tray. Output depends on the state only, that is, the FSM is a Moore machine.

Generators or *sequencers* are transducers with a single-letter input alphabet. They produce only one sequence of output.

In all the state machines considered so far, time is considered as a sequence of events only. The FSM does not specify how long the machine remains in a certain state. The time between two transitions that both require an input can be controlled by the time between the presentations of successive input symbols. If a sequence of transitions occurs with no input, no assumption can be made about how long the system remains in a state. Consider for example, a push-button traffic light at a pedestrian crossing. The sequence of states is obvious:

s_0 (cars: green, ped: red)

s_1 (car: yellow, ped: red)

s_3 (car: red, ped: red)

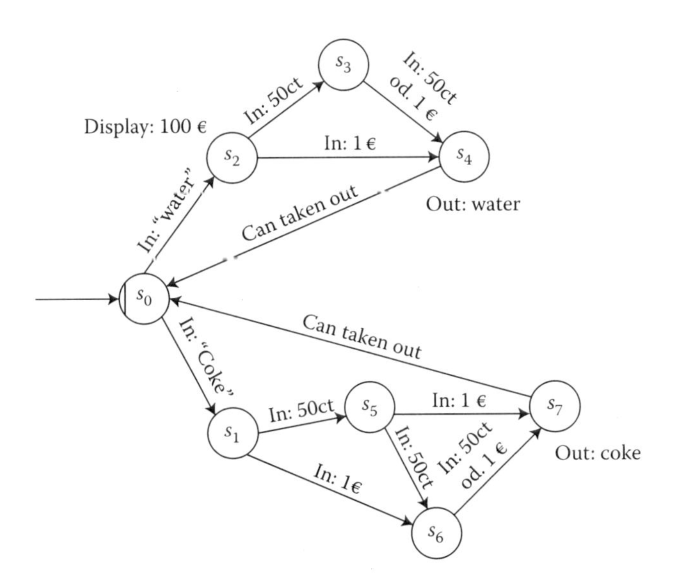

FIGURE 4.2
A vending machine.

s_4 (car: red; ped: green)

s_5 (car: red, ped: red)

s_6 (car: red + yellow, ped: red)

s_0

But there is no way to express that the traffic light should stay in state s_4 significantly longer than in s_3.

4.3.2.2 Timed Automata

FSMs abstract away from quantitative time and consider only sequences of events or states. This is referred to as the *linear time model*. When it comes to distributed real-time systems, however, a quantitative view of time is required. Timed automata model time quantitatively. In the linear time model, an execution trace of the system is just a sequence of states or events. Event *a* appears before event *b* in the trace if event *a* has happened earlier than event *b*. In the *discrete-time model*, time is modeled as a monotonically increasing sequence of integers. Events in an execution trace have time stamps. The *fictitious clock model* models time as a nondecreasing sequence of integers. It can be thought of as a synchronous system where events are sampled only at clock ticks. Although the discrete-time events may not occur at the same point of (modeled) time, the fictitious clock model allows multiple events to be sampled at the same clock tick. Timed execution traces of both models can easily be transformed into untimed execution traces of the linear time model, by inserting trace entries with no event (no state change) every clock tick.* In the models described so far, real-time is approximated by discrete points in time, that is, by integer values. Most physical processes, however, are described by a continuous view of time, for instance, by means of differential equations. The continuous nature of real time is reflected in the dense time model, where time stamps are real numbers. An execution trace in this model has a sequence of monotonically increasing time stamps. While integer-valued time can be modeled in the finite state machine framework by transforming timed execution traces to untimed execution traces, real-valued execution traces cannot be easily transformed into a form that can be dealt with by FSMs. Automatic reasoning about such a system is supported by the theory of timed automata. Henzinger [2] provides an introduction into this theory.

A timed automaton has a finite set of states, among them a start state and a set of final states and an input alphabet, like an FSM. Whereas in an FSM a sequence of transitions is triggered by the presentation of a word, that is, a sequence of symbols from the input alphabet, a sequence of transitions

* The discrete time model can be thought of as a fictitious clock model with a clock period of one time unit.

in a timed automaton is triggered by the presentation of a timed word (s,t), which is a sequence of pairs (s_i,t_i), where $s_i \in \mathbf{I}$ and $t_i \in \mathbb{R}$. The time sequence $t_1,t_2,t_i,t_{i+1}\ldots$ increases strictly monotonically, that is, $t_i < t_{i+1}$.

An alternative definition requires the time sequence to increase monotonically only, that is, $t_i \leq t_i + 1$, allowing multiple events (or transitions) to occur at the same time. Progress is guaranteed, that is, for every $t \in \mathbb{R}$, there is some $i \geq 1$ such that $t_i > t$.

For example, Figure 4.3 shows a timed automaton that describes a push-button traffic light at a pedestrian crossing.

It is initially in state (g,r), that is, green for cars, red for pedestrians. It is activated by pressing the button ("push" in Figure 4.3), which triggers the transition to state (y,r) and resets the clock c. At time t_{cy} the automaton changes to state $(r,r)_1$ with no further input. At time $t_2 = \Delta t_{cy} + \Delta t_{cpr}$, a transition to state (r,g) occurs. Δt_{cy} is the period of time during which the signal for the cars is yellow. Δt_{cpr} is the time during which the signals for both cars and pedestrians are red. Δt_{pg} is the duration of the green signal for pedestrians. At time $t_3 = t_2 + \Delta t_{pg}$, the pedestrian light turns to red; at $t_4 = t_3 + \Delta t_{pcr}$, the signal for cars goes to red and yellow; and at $t_5 = t_4 + \Delta t_{cry}$, the system returns to its initial state, that is, the car signal is green and the pedestrian signal is red. Without the time constraints, we would have an FSM that merely describes the sequence of states but does not specify how long the system remains in each state.

The t in the pair (s,t) can also be a time constraint, for instance $t > 10$. In our sample push-button traffic light control, we use a time constraint to guarantee a minimum "green" period of 10 time units to car traffic. To that end, we duplicate the state (g,r) and introduce a second clock, d, which is reset to zero at the end of the signaling sequence ($state(g,r)_2$). The transition back to the initial state ($state(g,r)_1$) has a time constraint of $d > 10$. As a result, the signal will stay green for cars for at least 10 units of time, even if the button has been pushed again, since the input will be read only if the initial state is reached again. Figure 4.4 shows the modified traffic light control.

The language of an FSM is the set of input words for which the FSM ends up in an accepting state. The class of languages that can be defined by an

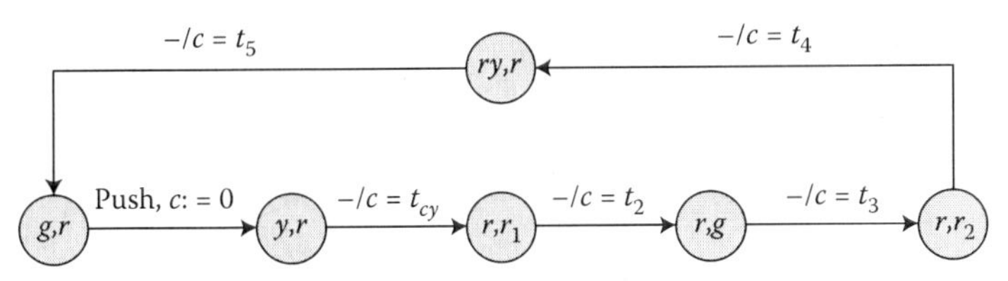

FIGURE 4.3
Traffic light.

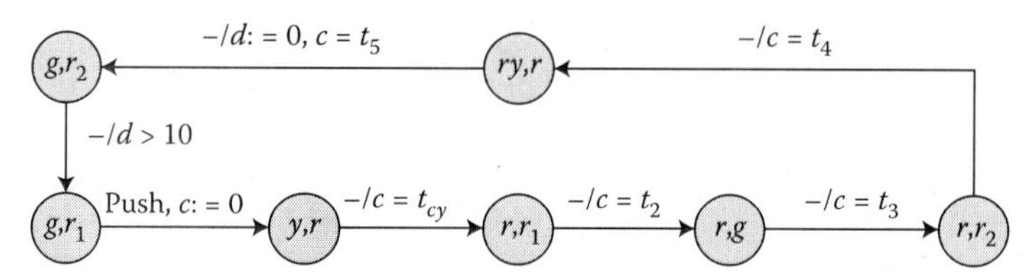

FIGURE 4.4
Modified traffic light control.

FSM is the set of regular expressions. The language of a timed automaton is the set of timed words for which the timed automaton ends up in an accepting state.

We will now illustrate the use of timed automata for verification with an example, which is taken from Alur and Dill [2] where the underlying theory is presented in detail.

Our example is a railroad crossing. We model the system with three components: train, gate, and controller. The system is modeled by the parallel execution of these three components. Figure 4.5 shows the automata for this example.

In Figure 4.5 id_T is the train's idling event, which states that the train is not required to pass the crossing. The train generates an event *approach* when it is at a certain distance of the crossing. The event could be implemented by a sensor on the track that sends a signal. Events *in* and *out* are generated when the train enters and leaves the crossing. Event *exit* is generated when the train is at a certain distance away from the crossing. The process *train* communicates with the process *controller* using the events *approach* and *exit*. The time constraint $(x > 2)$? states that the train must generate the event *approach* at least 2 minutes before it reaches the crossing. The time constraint $(x < 5)$? associated with event *exit* states that the train will leave the region around the crossing no later than 5 minutes after the event *approach* has occurred. This is a *liveness* requirement.

The gate has the following events: *raise* (start to open the gate), *lower* (start to close the gate), *up* (has reached the open position), and *down* (has reached the closed position). It may remain in the open or closed position (states s_0 and s_2) for an arbitrary amount of time (event id_G). These states, therefore, are both initial states. The process of closing the gate takes at most 1 minute, starting with event *lower* in the open position and reaching the closed position with event *down*. The gate responds to the *raise* event by reaching the open position between 1 and 2 minutes.

The controller has the initial state s_0, which is also its idle state. If it receives an *approach* signal from the train, it will send a *lower* signal to the gate within

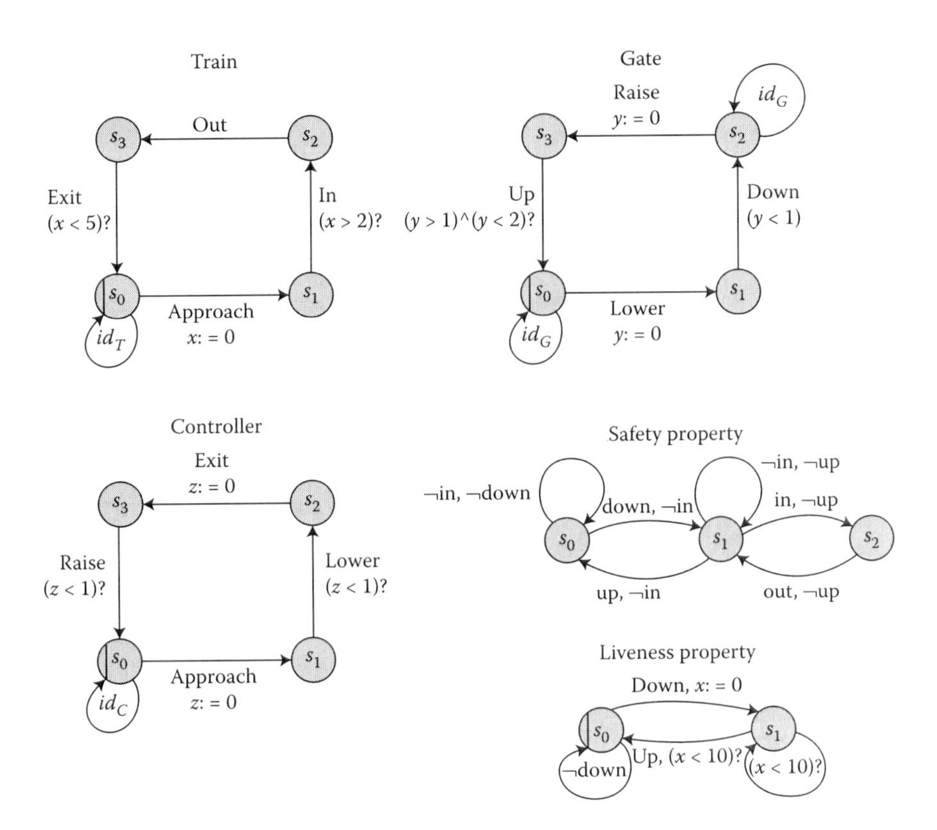

FIGURE 4.5
Modeling a rail-gate controller.

1 minute. Upon receipt of an *exit* signal from the train, the controller sends a *raise* signal to the gate within 1 minute.

Timed automata can also be used to express important properties of the system:

Safety—Whenever the train is in the crossing, the gate must be closed.

Real-time liveness—The gate will open after at most 10 minutes.

The specification uses only events *in*, *out*, *up*, and *down*. An edge label "*in*, *up*" means any event set containing *in* but not *up*. In the safety automaton, all states are accepting states. The automaton states that *down* must occur before *in* and *up* must occur after *out*, that is, the gate must be closed when a train enters the crossing and must not be open until the train has left the crossing.

The liveness automaton will not stay in state s_1 for more than 10 minutes. In state s_1 the gate is closed. s_1 is reached from initial state s_0 by receiving a *down* signal. After at most 10 minutes an *up* event will occur that takes the system back to state s_0, where it stays as long as no *down* event is received.

4.3.2.3 Hybrid Automata

Hybrid automata are a generalization of timed automata. Whereas in timed automata, only time is modeled as a real-valued quantity, in hybrid automata multiple variables are real-valued and their behavior is described by differential equations.

A hybrid automaton **H** has the following components:

- A finite set $\mathbf{X} = \{x_1, x_2, \ldots, x_n\}$ of real-valued variables, that is, $x_i \in \mathbb{R}$ Their number, n, is called the dimension of the automaton **H**. The set of the first derivatives during continuous change of the variables is represented by $\dot{\mathbf{X}} = \{\dot{x}_1, \dot{x}_2 \ldots \dot{x}_n\}$. Primed variables $\mathbf{X}' = \{x_1', x_2' \ldots x_n'\}$ denote the values at the conclusion of discrete change.
- A *control graph*, that is, a directed multigraph—Its vertices are the *control modes*, and its edges are called *control switches*. A multigraph can have multiple edges that have the same end nodes.
- Initial, invariant, and flow conditions are associated with every control mode v. The vertex labeling functions *init*, *inv*, and *flow* assign predicates to each control mode.
    - An initial condition *init*(v) is a predicate whose free variables are from **X**.
    - An invariant condition *inv*(v) is a predicate whose free variables are from **X**.
    - A flow condition flow(v) is a predicate whose free variables are from $\mathbf{X} \cup \dot{\mathbf{X}}$.
- Jump conditions—An edge labeling function *jump* assigns to each control switch $e \in \mathbf{E}$ a predicate *jump* (e) whose free variables are from $\mathbf{X} \cup \dot{\mathbf{X}}$.
- Events—A finite set Σ of events and an edge labeling function event; $E \rightarrow \Sigma$ assigns an event to each control switch.

The following example illustrates the preceding definition. The definition and the example are taken from Henzinger [3], which gives a detailed introduction to the theory of hybrid automata.

The hybrid automation shown in Figure 4.6 models a thermostat.

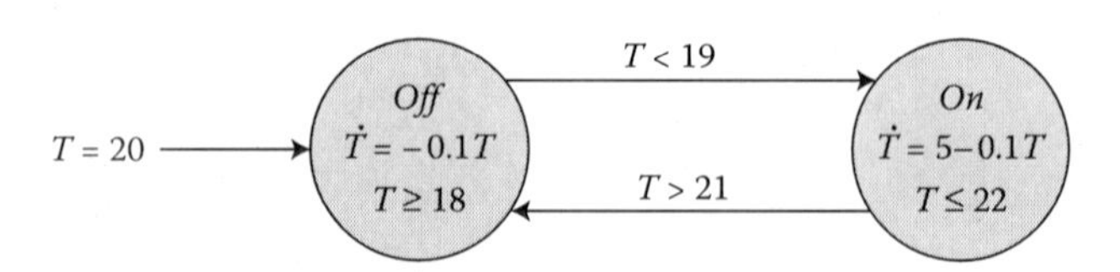

FIGURE 4.6
Hybrid automata representation of a thermostat.

The variable θ represents the temperature. In control mode *Off*, the heater is inactive and the temperature falls as described by the flow condition

$$flow(Off): \dot{\theta} = -0.1\theta$$

In control mode *On*, the temperature rises as

$$flow(On): \dot{\theta} = 5 - 0.1\theta$$

The initial state of the system is

$$\text{heater off}, \quad \theta = 20°\text{C}$$

The jump condition $\theta < 19°C$ specifies that the heater may go on as soon as the temperature falls below 19°C. The invariant $\theta \geq 18°C$ says that the heater will go on when the temperature falls below 18°C. Similarly, the heater may go off if the temperature rises beyond 21°C (jump condition), and it will go off if the temperature goes above 22°C (invariant).

4.4 MATLAB Stateflow

MATLAB [4] Stateflow charts can be used to represent event-driven systems using sequential logic based on FSMs. A Stateflow chart is a graphical representation of an FSM. The *states* and *transitions* form the basic building blocks of the sequential logic system used by Stateflow.

The basic module for building a model using Stateflow is a *chart*, which can be included in a Simulink® model like any other block from the Simulink Library. A chart allows a developer to define an FSM using graphical blocks like states, transitions, conditions, actions, input–output data, and events. Details of MATLAB Stateflow are available in Reference 5. In this section the Stateflow representation of an FSM is illustrated with a simple example: the vending machine represented in Figure 4.2 as a Moore machine.

With the FSM defined by Figure 4.2, the corresponding Stateflow implementation is represented by the chart shown in Figure 4.7.

In Figure 4.7, the states are represented by the rectangular boxes and the transitions by the directed lines connecting two states. The condition for each transition is enclosed between two square brackets [], and the corresponding transition equation, which is executed once the condition is evaluated to be true, is enclosed between two curly brackets {}. The variables *coin*, *op*, and *w* are inputs to the chart representing, respectively, the denomination of the coin (i.e., 1€ or 0.5€); the selected option, that is, 1 for coke and 2 for water; and a flag indicating that the user is collecting the bottle. Similarly the variables

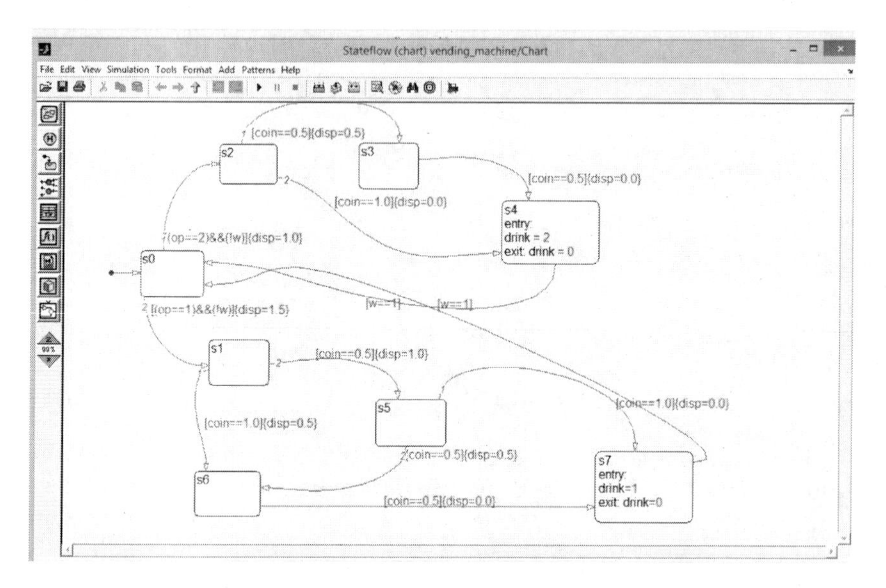

FIGURE 4.7
Representation of a vending machine using MATLAB Stateflow.

disp and *drink* denote the pending amount and the drink dispensed, which is again 1 for coke and 2 for water. The directives *entry* and *exit* denote the actions taken by the chart when a particular state is entered and when an exit occurs, respectively. Similarly, the directive *during* can be used in states s_4 and s_7 to activate a function *dispense_bottle*() to actually dispense the bottle containing the selected drink. This is not shown in the Stateflow chart and is left as an exercise for the reader. Another implicit assumption is this: the machine does not accept a fresh input until the present user collects the bottle when the flag w is reset.

The functioning of this chart can be understood by analyzing the corresponding Simulink model represented in Figure 4.8.

The model shows a case where the user selects the option $op = 2$, that is, water, and inserts a 1€ coin. With this, the activated state in Figure 4.7 becomes s_4, as shown in Figure 4.9.

The state s_4 shall remain activated until the user withdraws the bottle, which is accomplished by clicking on the *manual switch*, shown in Figure 4.8, to move it from 0 to 1, thus changing the variable w accordingly. This causes a transition from state s_4 to s_0, as shown in Figure 4.10.

It is clearly seen from an examination of Figures 4.7 through 4.10 that transitions from state s_0 remain blocked until the flag is reset by switching back the manual switch to position 0, which indicates that the user has completed the transaction.

The chart shown in Figure 4.7 is very basic and can be implemented in a variety of ways using more options detailed in Reference 5.

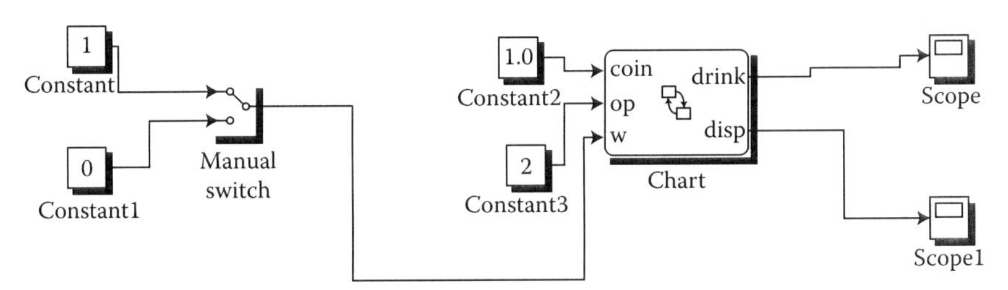

FIGURE 4.8
Simulink model for a vending machine.

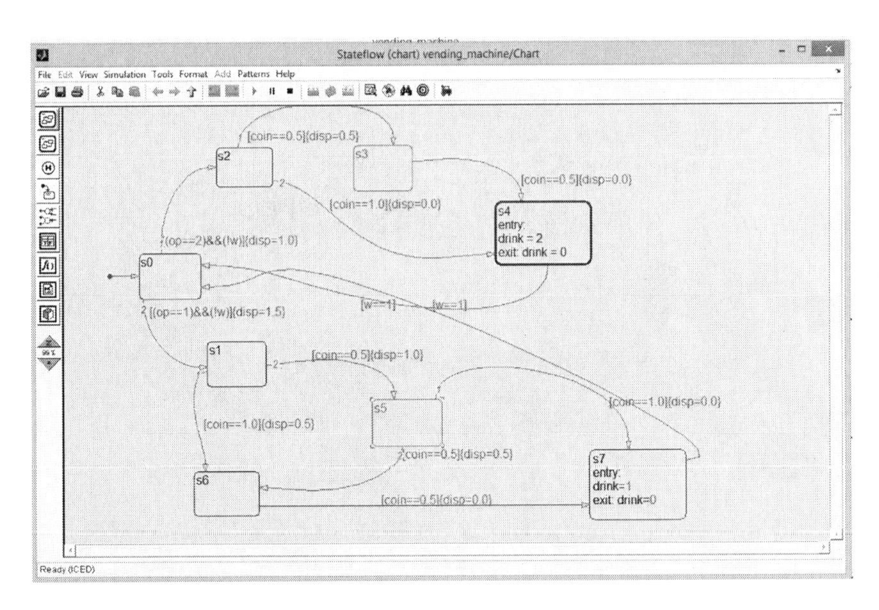

FIGURE 4.9
(See color insert.) State activation in the chart of Figure 4.7 corresponding to the model in Figure 4.8.

NUMERICAL AND ANALYTICAL PROBLEMS

4.1 A full-adder circuit accepts two bits X_1, X_2 as inputs and produces an output bit Y and a carry bit C. Represent this as (a) a Mealy machine and (b) a Moore machine. Which representation requires fewer states?

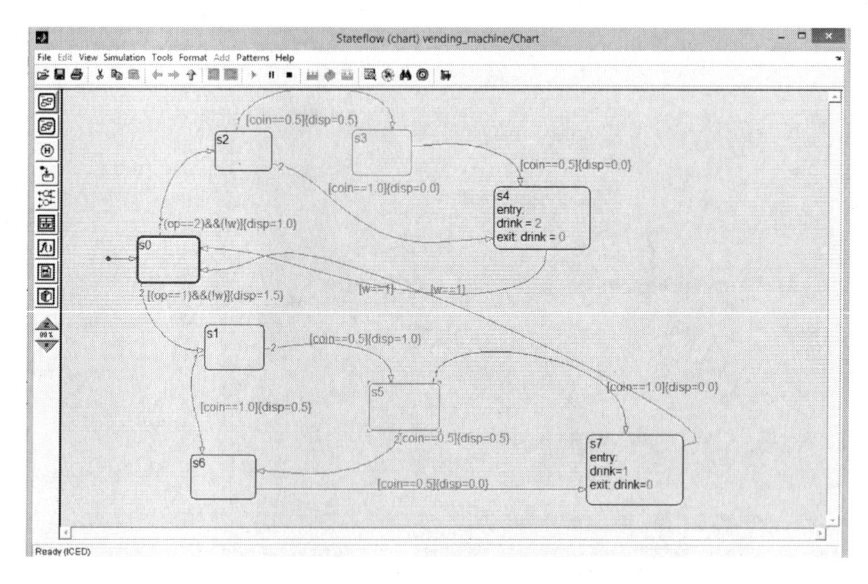

FIGURE 4.10
(**See color insert.**) State activation in the chart of Figure 4.7 corresponding to the model in Figure 4.8 during retraction of bottle.

(a)

(b)

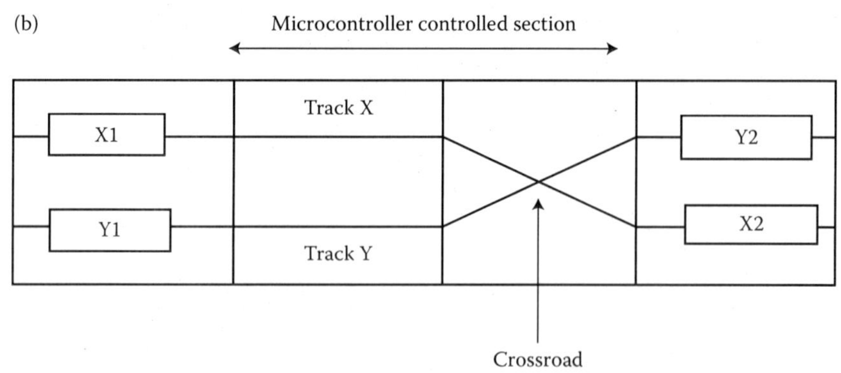

FIGURE 4.11
(**See color insert.**) (a) Track with crossroad. (b) Sensors on crossroad.

4.2 Represent the traffic light controller of Figure 4.1 as a Stateflow chart using a suitable representation of the FSM, that is, a Mealy or a Moore machine.

4.3 Augment the model of Figure 4.8 by adding another Stateflow block that uses another chart to represent the dispenser using the variable drink as the input.

4.4 Consider a toy comprising two cars that run on electric tracks, as shown in the Figure 4.11a [6]. Two toy cars ply on the two tracks X and Y, and the speed is controlled by varying the current in the tracks using a microcontroller. Each track has one pair of sensors to monitor approach and recession of the cars relative to the crossroad (Figure 4.11b). Model the system as (a) an event-triggered system and (b) a time-triggered system and represent the control task using MATLAB Stateflow so that car on the X track stops when both cars approach the crossroad simultaneously.

References

1. Kopetz, H. *Real-Time Systems: Design Principles for Distributed Embedded Applications*. Springer, 2011, 978–144, 198, 236.
2. Alur, R. and Dill, D. L. A theory of timed automata. *Theoretical Computer Science*, 126(2), 183–235, 1994.
3. Henzinger, T. A. The theory of hybrid automata, in *Verification of Digital and Hybrid Systems*, eds. M. K. Inan and R. P. Kurshan. Springer (NATO ASI series), 2000, 265–292.
4. MATLAB. www.mathworks.com.
5. MathWorks. Stateflow User's Guide. http://nl.mathworks.com/help/pdf_doc/stateflow/sf_ug.pdf.
6. Mandal, T. Development of a microcontroller based Automatic traffic control system using slot cars. Master's dissertation, Jadavpur University, 2013. http://dspace.jdvu.ac.in/handle/123,456,789/28,411?mode – full.

5

Developing Distributed Real-Time System Applications—the MATLAB Way

MATLAB is an effective tool for simulation of dynamic systems. Though mostly used for simulation using a simulation timeline, MATLAB can also be used for simulation of dynamic systems on a real-time basis, that is, using a real clock as the timeline for simulation. Simulations can be run within the MATLAB environment as SIMULINK® models or as stand-alone systems, which can be used to develop distributed real-time system (DRTS) applications. MATLAB can also be used to for simulation of DRTS applications within the MATLAB environment. These methodologies are discussed in this chapter.

5.1 Developing MATLAB Real-Time Targets

The Real-Time Workshop (RTWS) is the basic module of MATLAB, which is used for generating real-time code from the model designed with the help of modules like the Simulink, Stateflow and xPC, for running either as a stand-alone application or within the MATLAB environment. It provides options for generating more efficient code with the help of the RTWS Embedded Coder (ERT) to produce more compact and faster code than RTWS. It also provides built-in support for software standards such as AUTOSAR (standard used in automotive software). For building an executable from a model a *make file* is required, which provides various configurations for controlling the compilation and target specification. A Target Language Compiler (TLC) is used for generating code for the target system.

Figure 5.1 shows the general schematic of code development using the RTWS.

Using the RTWS assumes that a C++ compiler is installed in the system and the most used one is a Microsoft Visual C++. For running a model in real-time within the MATLAB environment using Simulink, the *Real-Time Windows Target* must be installed using the *rtwintgt–setup* command at the MATLAB command prompt, details of which are available in the MATLAB documentation.

Figure 5.2 shows a Simulink model of a real-time task that generates a square wave of amplitude 2 volt with a time period of 1 s and a duty-cycle of 0.5 using an analog output card PCI-1711U [1] and the corresponding output

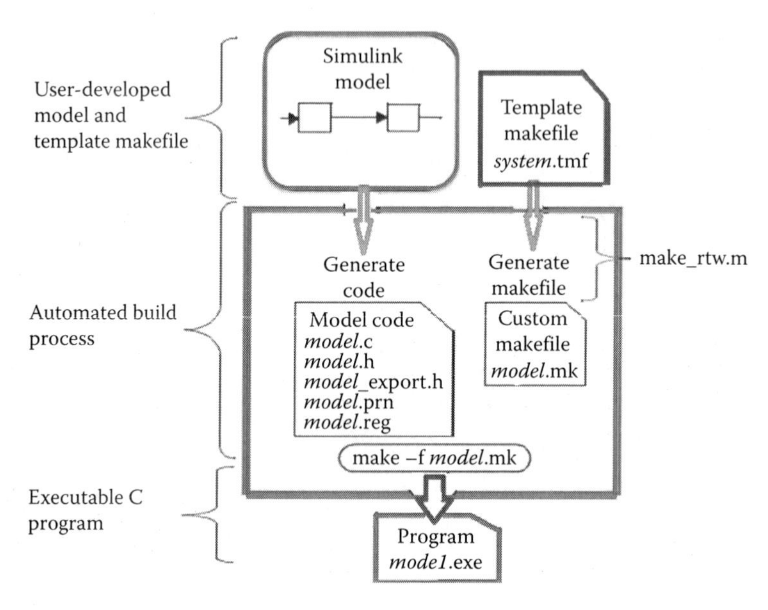

FIGURE 5.1
Building an RTWS application.

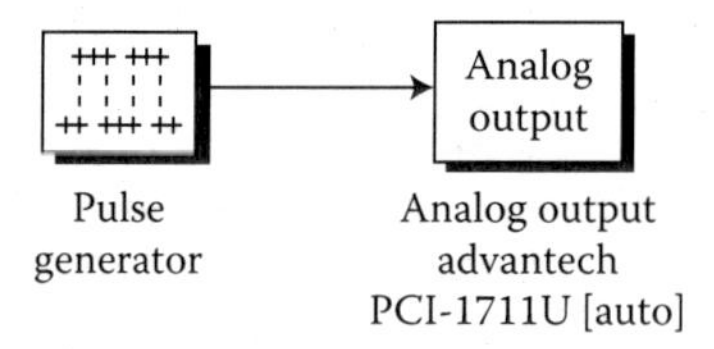

FIGURE 5.2
Simulink model of a real-time task for execution in a MATLAB environment.

captured on an oscilloscope as shown in Figure 5.3. The trigger is chosen as external and the corresponding target language compiler is *rtwin.tlc*.

The square-wave source is chosen as a *sample-based* one and the time-step of simulation chosen is 100 ms. The Target needs to be connected first using the button next to the simulation time on the model pane and then run in the usual way.

5.2 Using the xPC Target

The MATLAB toolbox xPC Target [2] provides a real-time rapid-prototyping tool on a Host-Target environment in conjunction with MATLAB and other

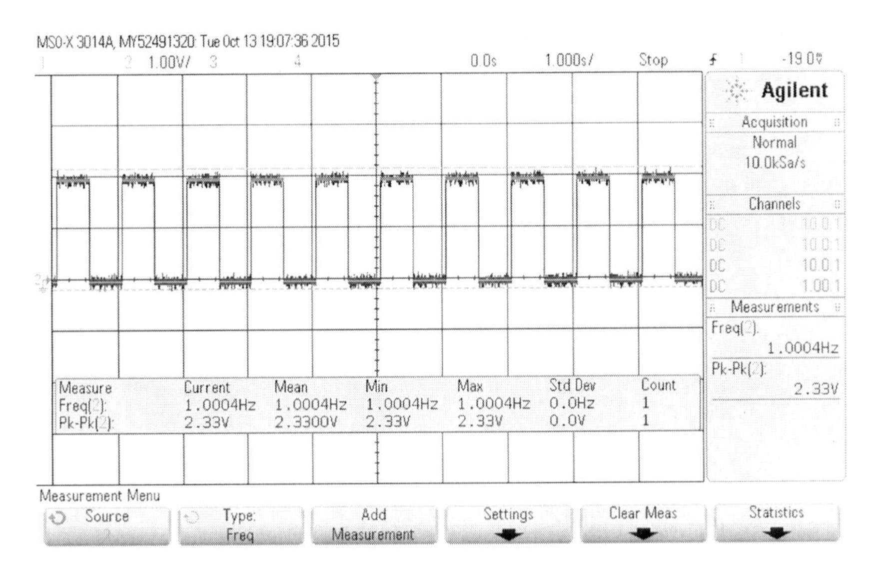

FIGURE 5.3
Output of the model in Figure 5.2 on an oscilloscope.

toolboxes to model and design real-time control systems and implement in a hardware-in-loop (HIL) simulation. The general arrangement is shown in Figure 5.4.

The *Target* is the system in which the embedded application runs. This usually contains a bootable image of the xPC Target kernel or must be booted with one before connectivity with the host is established. The *host* is a standard microcomputer, that is, a desktop, laptop or a workstation with an OS like windows or Linux on which MATLAB runs.

The xPC Target module on the host acts as an integrated development environment (IDE) to develop embedded applications that run on the target.

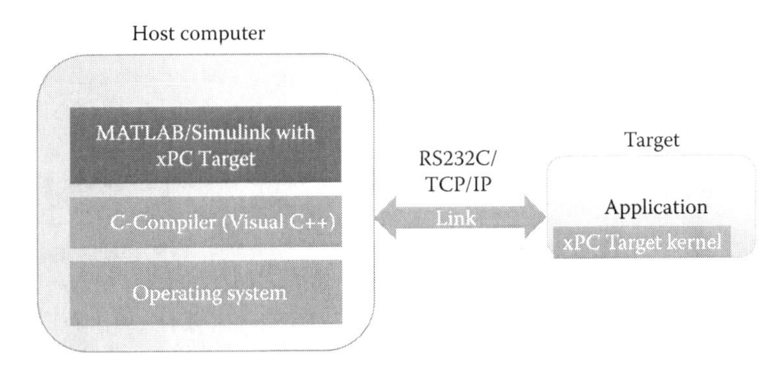

FIGURE 5.4
xPC Host–Target system.

FIGURE 5.5
The xPC Target module.

The embedded application can be developed using Simulink and other blocks that are listed in the xPC Target module, as shown in Figure 5.5.

A guide to hardware supported by the xPC Target utility can be found online [2]. In general, the IDE supports Intel, AMD, and other x86 compatible platforms and includes drivers for a wide variety of PC-compatible peripherals. The compiler is usually a visual C++. The host computer is configured for developing and downloading target applications using the *xpcexplr* utility. The utility is selected by using the command *xpcexplr* at the MATLAB prompt, which opens the window shown in Figure 5.6.

The essential steps in configuring the host system for downloading an application are as follows:

1. *Setting up the C compiler*—The easiest way to do this is by clicking on the compiler configuration tab and then the tabs on the window shown in Figure 5.7.

   If Visual C++ is already installed, the compiler path needs to be set for the xPC Target to be able to build executable applications.

2. *Configuring Target and Target Interface*—This tells the IDE about the target hardware and the communication interface. The targets can be added using the toolbar on the *xpcexplr* window. In case multiple targets are present, the default can be set by a right-click on the Target name. Targets can also be renamed in the same way. The targets are configured by selecting the appropriate properties, for

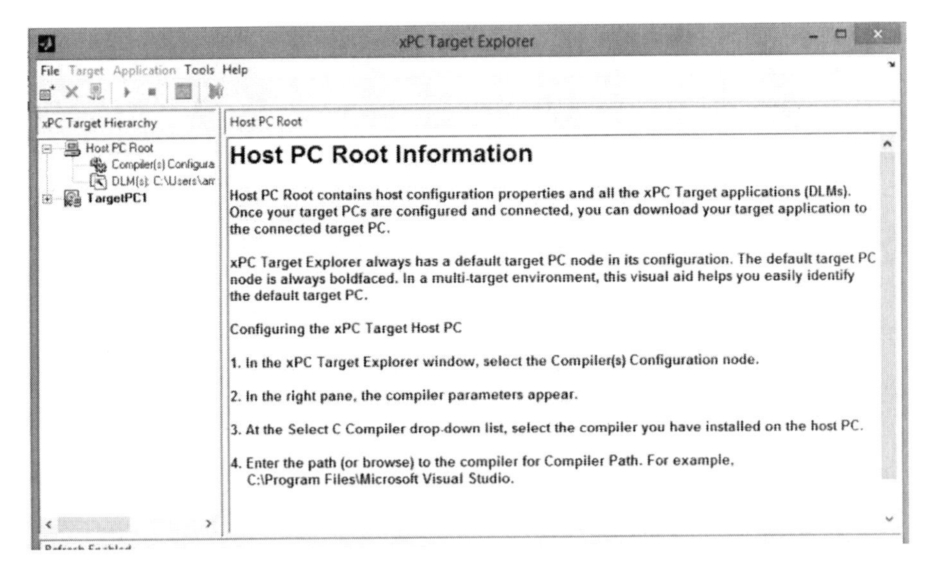

FIGURE 5.6
Configuring the host using *xpcexplr*.

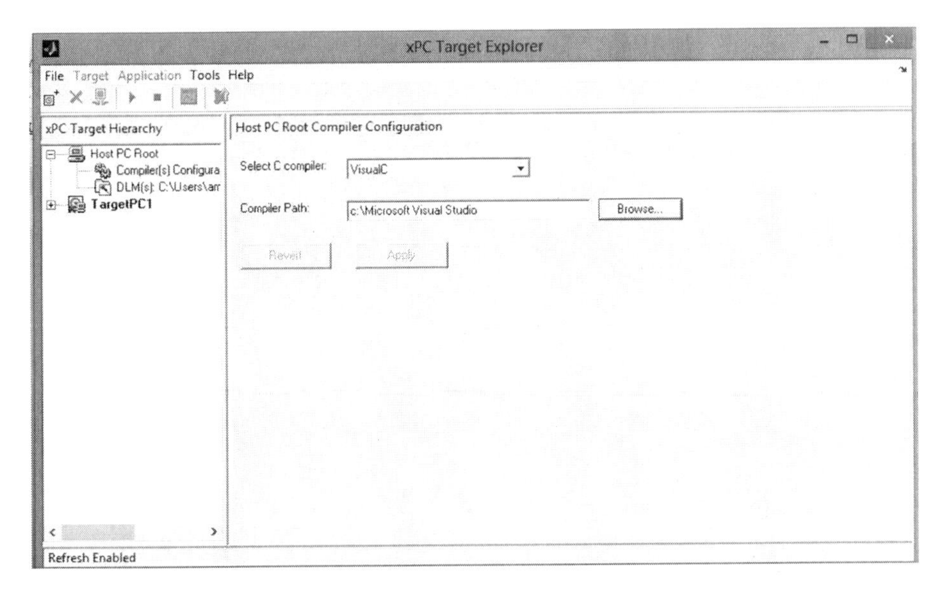

FIGURE 5.7
Setting up the compiler path on xPC Host.

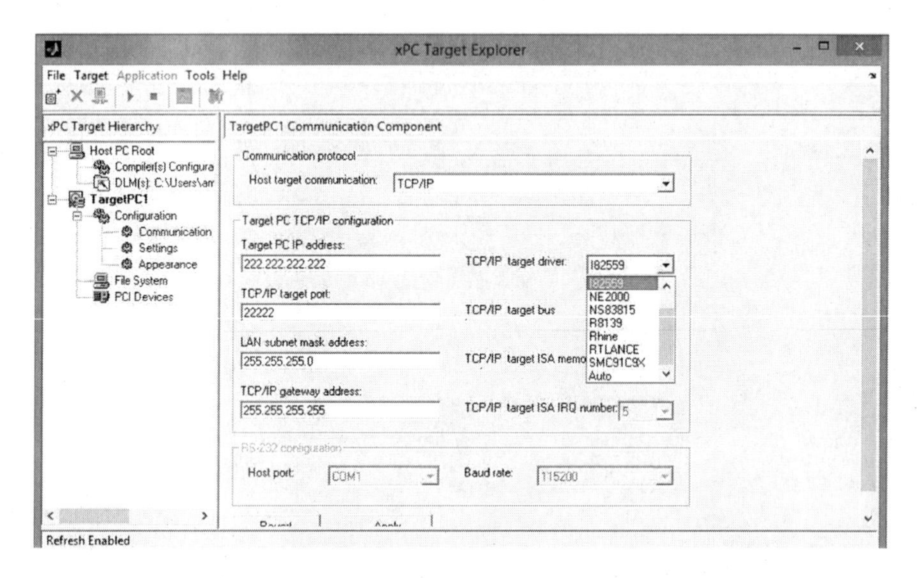

FIGURE 5.8
Configuring Target interface on xPC Host.

example, *Communications*, which is selected for TargetPC1 (default nomenclature) as shown in Figure 5.8.

As shown in Figure 5.8, the xPC Host is set up to communicate with the TargetPC1 through a TCP/IP interface, with options for network interface and the properties relevant for TCP/IP as shown in the window. The specification for the network interface is required so that the appropriate driver is bundled with the application. The IDE selects the default target for downloading the application identified through the IP address.

Once the application is downloaded on a target, the target has to be selected by left-click and then connected to the host using the *Target > Connect to Target* option on the toolbar of the xPC Target Explorer window. The application is started by clicking on the play button.

Figure 5.9 shows xPC implementation of a Target *Send-Receive*, which sends a sine wave over a UDP/IP interface to the Target *Reflector* (Figure 5.10), and the sent and received sine waves captured on the *Target Scope* of the Send-Receive module is reproduced in Figure 5.11 using the *xpctargetspy* utility. A detailed description of the usage of the xPC blocks used in the models of Figures 5.9 and 5.10 as well as the *xpctargetspy* is available in a guidebook [2].

As seen in Figure 5.10, the *Reflector* sends back the transmitted sine wave to the Send-Receive module and there is a minimum time delay, which equals one sample time between the transmitted and the received signals seen by the Send-Receive module that causes a small phase lag in the received (or reflected) sine wave. The initial distortion in the reflected wave is due to the

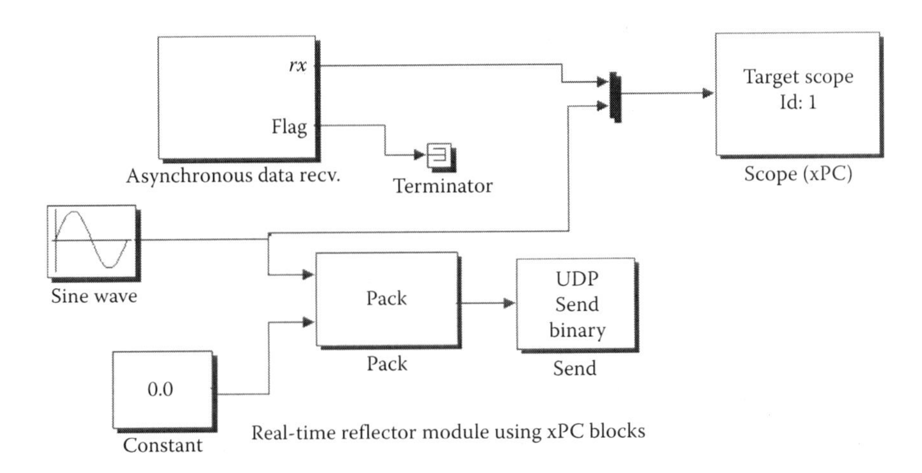

FIGURE 5.9
xPC Target implementation of a real-time Send-Receive module.

substantial data loss caused by the traffic generated by the refresh option by the host computer, which is then disabled using the Disable Refresh option by clicking on the Tools tab on the *xpcexplr* pane, thus reducing the data loss in the reflected wave.

It is important to note that building xPC Targets requires proper choice of the Target Language Compiler and a fixed step size. While choice of compilers allows various level of optimizations, the simple choice is *xpctarget.tlc,* which is selectable using the configuration option from the simulation tab

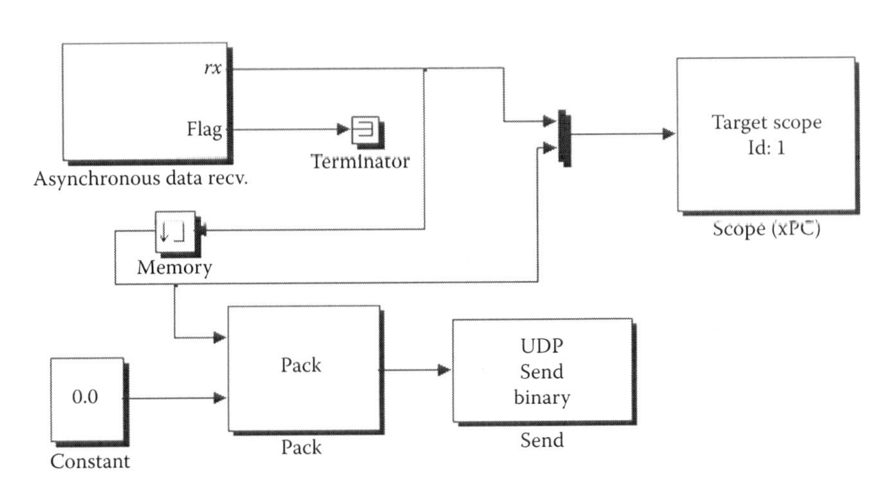

FIGURE 5.10
xPC Target implementation of a real-time Reflector module.

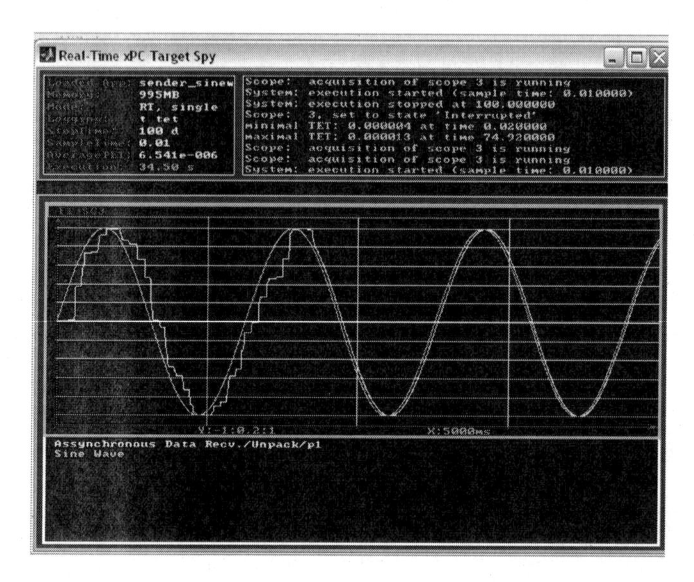

FIGURE 5.11
(**See color insert.**) The transmitted and received sine waves on the Send-Receive module.

in the pane of the Simulink model. Using the *build* or the *incremental build* option in the configuration pane builds the Target and downloads it on the default target selected through *xpcexplr.*

5.3 Building a Distributed Real-Time Application: Simulation of Communication between Earth, a Lunar Orbiter, and a Lunar Lander

The scenario considered is an unmanned space mission, for example, to the moon or to mars, which has the following components: *earth station, an orbiter* in a lunar orbit, and a *rover.* Figure 5.12 illustrates the scenario.

An orbiter is a lunar satellite orbiting the moon at a fixed radius around a polar orbit so that it always remains visible from the Earth. A *lander* or a rover is dispatched on the lunar surface to perform physical exploration. The rover is usually controlled through the orbiter from an earth station located on Earth. As evident from Figure 5.12, the communication between the various components of the earth–orbiter–rover system involves varying latency and at times the components may not be visible. Communication in such deep-space missions is usually done using a specialized protocol like Proximity-1, which has been explained in Chapter 3. This section attempts to develop a simulation platform that can be used

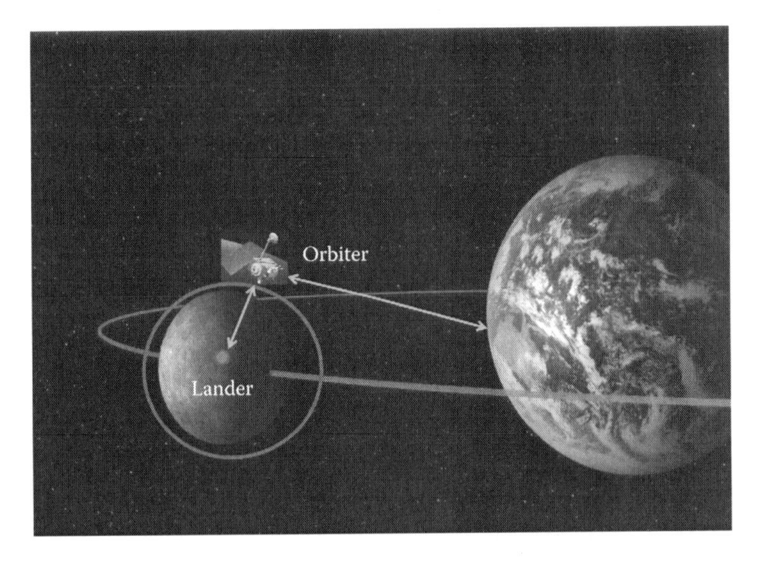

FIGURE 5.12
(**See color insert.**) Communication between earth station, orbiter, and rover.

to simulate varying delay in this deep-space communication scenario so that performance of control algorithms to manage activities such as descent of the lander or rover and the trajectory of the orbiter can be evaluated [3]. The same framework can also be used to evaluate activities like instrument calibration and real-time system updates. The simulation uses MATLAB Stateflow and xPC Target.

The DRTS application of Figure 5.12 involves four real-time simulation modules connected over UDP/IP. Each simulation module consists of a communication layer with an interface for the application. The communication module is time-triggered, and messages to and from the respective applications are queued until a receive or send is scheduled. Three of the simulation modules represent the earth station, the orbiter, and the lander/rover, and the fourth is defined as an *event scheduler*, which acts as an omniscient observer that intercepts all messages and dispatches them to the recipients after adding a requisite delay from visibility and distance considerations between the source and the destination. Transmission of every message denotes an event and every event is time-stamped according to a global clock. For the present system, the event-scheduler acts as the master for clock synchronization.

The communication module in each of the three modules (i.e., the earth station, the orbiter, and the rover) can be represented by a general Moore machine template shown in Figure 5.13a, while the event scheduler can be represented by Figure 5.13b. In Figure 5.13a and b the subscripts e, o, r, and s represent the earth station, orbiter, rover, and event scheduler, respectively, and the subscript $\alpha \in [e, o, r]$ in Figure 5.13a is used to represent a general template of control flow.

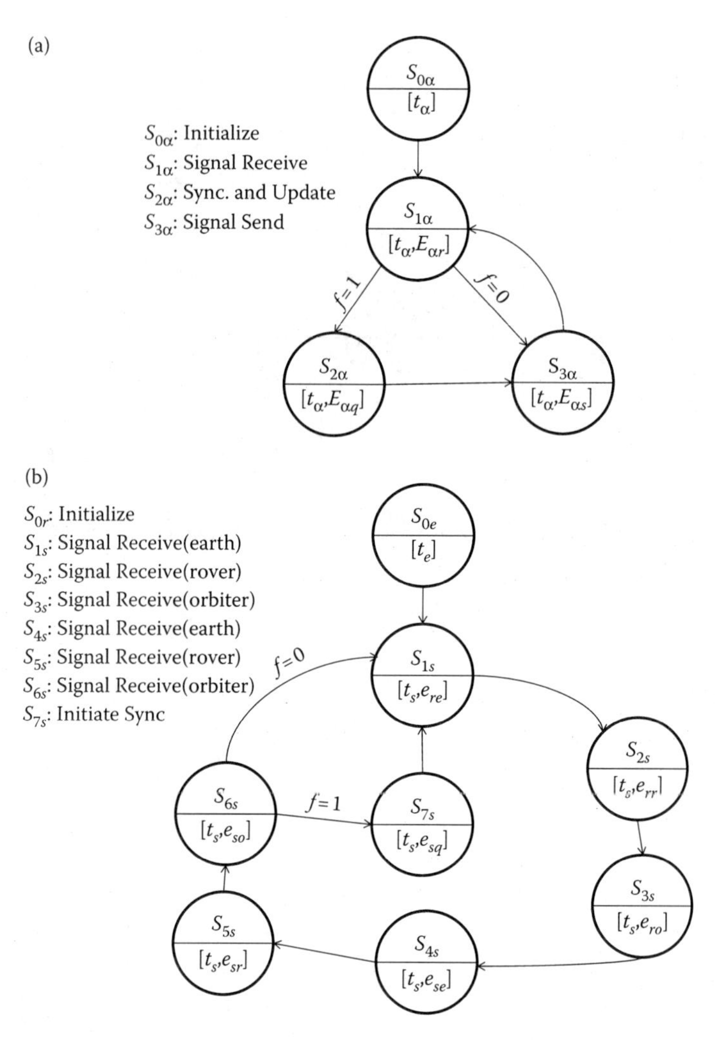

FIGURE 5.13
(a) Moore machine template. (b) Moore machine template for earth station, orbiter, and rover for the event scheduler.

For the finite state machines (FSMs) in Figure 5.13, the input is simply a flag f, $f = 1$ indicates that a new time synchronization data is available for the earth station, orbiter, and the rover, and denotes that time synchronization is due for initiation by the event scheduler module.

Similarly, the corresponding output alphabets for the different Moore machines in Figure 5.13 are defined as follows:

$$\Lambda_\alpha = [t_\alpha, E_{ar}, E_{as}, E_{aq}]\ \alpha \in [e, o, r]$$
$$\Lambda_s = [t_s, e_{er}, e_{or}, e_{rr}, e_{es}, e_{os}, e_{rs}, e_{sq}] \tag{5.1}$$

The variable t_α, $\alpha \in [e, o, r]$ denotes the local time recorded at the earth station, the orbiter, and the rover, respectively. The variable t_s denotes the time recorded by the clock on the event scheduler, which acts as the master clock for the simulation. An event $E_{\alpha r}$ is output from the corresponding Moore machine to process any new message from the event scheduler. Similarly, an output event $E_{\alpha s}$ is used to sequence sending of a queued message from the corresponding simulation module to the event scheduler. A message has the general format $\{\alpha, \beta, t_\alpha, L, \mathbf{D}\}$ where $\alpha, \beta, \in [e, o, r]$ denote the source and destination, respectively; t_α denotes the time recorded by the local clock on the source at the instant the message is sent; and L denotes the number of data elements in the payload $\mathbf{D} \subset \mathbb{R}$. The event scheduler regularly outputs events $e_{\alpha r}$, $\alpha \in [e, o, r]$ to process messages—a new arrival is parsed to decode its destination, the delay is calculated based on the distance and visibility between the source and the destination, and the message is queued in the send queue for its destination with a delay counter, which is decremented by unity at every subsequent time-step.

The output events $e_{\alpha s}$, $\alpha \in [e, o, r]$ are similarly used by the event scheduler to check the message queues of the destinations $\beta \in [e, o, r]$ for all queued messages for which the counter has elapsed, but while sending the message, the time stamp of the original message is kept unaltered. Thus, the message is received at the destination in real-time with the delay it would have actually seen in a real deep-space communication scenario.

The flag f is set in the event scheduler module at the expiry of every resynchronization interval R, which is user defined causing an appropriate transition in the FSM representing the event scheduler with an output $[t_s, e_{sq}]$. The event e_{sq} causes the application on the event scheduler to send a message instantly with payload $\mathbf{D} = [0.0]$ and the time stamp t_s for every destination $\beta \in [e, o, r]$. The time-synchronization messages are sent on a different dedicated port on each simulation module, which is polled by the application in response to the output event $E_{\alpha q}$ generated by the corresponding FSM representing its communication module. Figure 5.14 represents the Simulink model for the rover following the general template represented by Figure 5.13a.

The model in Figure 5.14 assumes that the sending of messages is synchronized with the control flow of the communication module and all queued messages are sent to the event scheduler. However, the receipt of messages is assumed to be asynchronous. Each message received over the network generates a pulse, the rising edge of which causes the data to be latched and sets a flag indicating arrival of new data. This flag is reset only when the data is read-either in response to the event *eq* or *er* in the Simulink model of Figure 5.14, which correspond to events $E_{\alpha q}$ and $E_{\alpha r}$ in the Moore machine of Figure 5.13a for $\alpha = r$. The asynchronous receiver is handled by the Receiver Ctrl. blocks in Figure 5.14, represented using MATLAB Stateflow in Figure 5.15.

The chart in Figure 5.15 is programmed to be executed on initialization and a wake up is caused by an event *ea* or *eb*, which are input to the chart

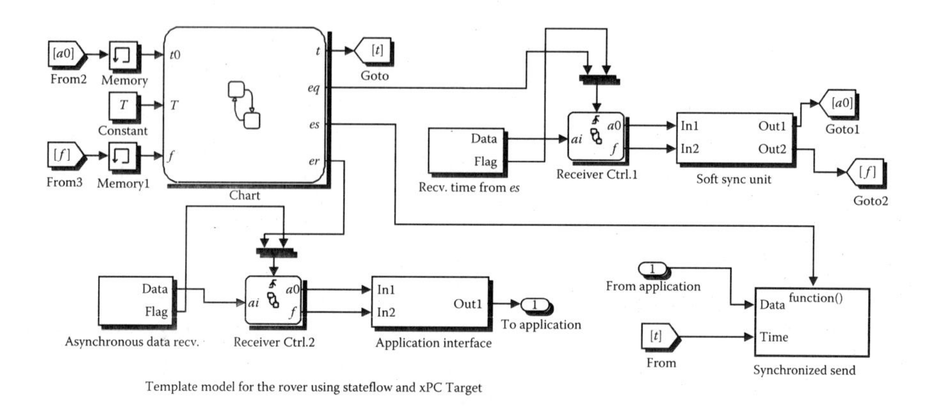

FIGURE 5.14
Simulink model of the communication module on the rover.

and are generated when fresh data is received over the network and when the data is read by the application, respectively. These inputs to the chart of Figure 5.15 are programmed in such a manner that if both events occur at the same instant, the event *ea* causes the wake up. The events *ea* and *eb* are generated when the signals connected to them undergo a transition from one level to another and the chart of Figure 5.15 is programmed for activation on a *rising edge*. Occurrence of the event *ea* means receipt of new data, which causes the input variable *ai* to be transferred to the output *a*0 and the flag *f* to be set. The subsequent activation of the chart by the event *eb* means that the application has read the data and the flag *f* is now reset. Figure 5.15 shows the

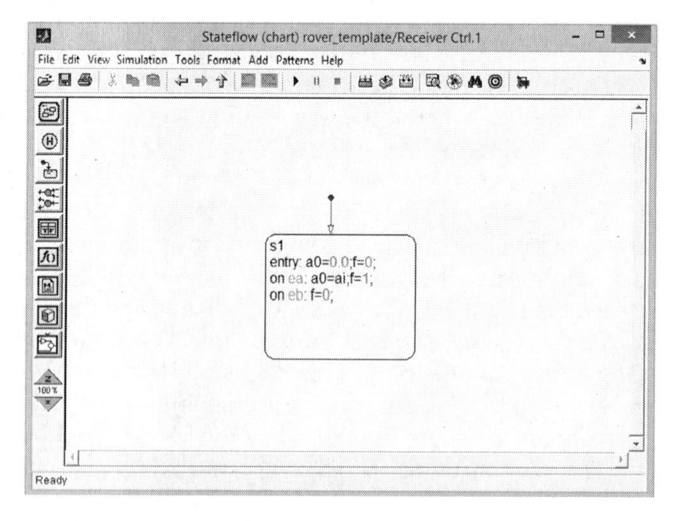

FIGURE 5.15
Stateflow representation of Receiver Ctrl. blocks shown in Figure 5.13.

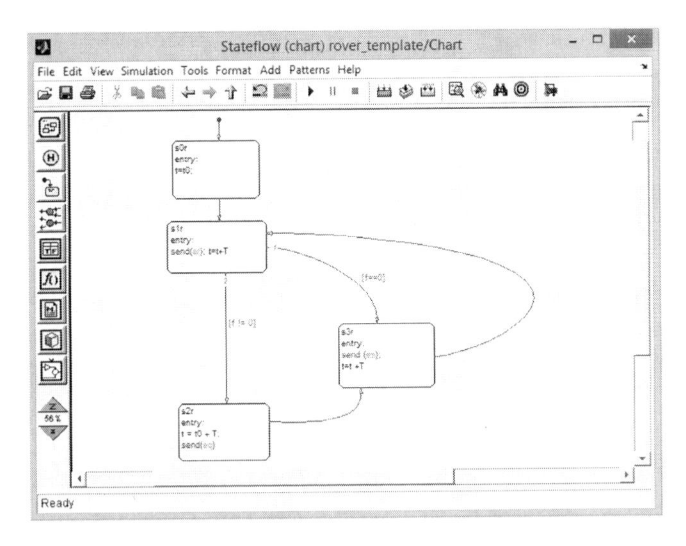

FIGURE 5.16
Stateflow chart representing the communication control module on the rover.

general template of a block that can be used to receive asynchronous time-synchronization and application data through the *Receiver Ctrl. 1* block or the *Receiver Ctrl. 2* block as shown in Figure 5.14.

The control of the communication module for the rover shown in Figure 5.14 is achieved by the *Chart* block, which is shown in Figure 5.16. The Stateflow chart in Figure 5.16 is an exact implementation of the Moore machine in Figure 5.13a. The events *er* and *eq* are synonymous with the events *eb* in the blocks *Receiver Ctrl. 2* and the *Receiver Ctrl. 1*, respectively, and are used to query arrival of fresh application and time-synchronization data from the event scheduler, respectively. However, while the flag *f* from the *Receiver Ctrl. 1* block is used for time synchronization and hence is an input for the control block represented by *Chart* in Figure 5.14, the output flag *f* from the *Receiver Ctrl. 2* block is used by the application on the rover module to transfer new data from the event scheduler. The blocks *Receiver Ctrl. 1* and *Receiver Ctrl. 2* are used provide an asynchronous interface for the time-triggered structure of the communication control module represented by the *Chart* block in Figure 5.14. However, for sending of messages in Figure 5.14 the block *Synchronized send* is active only when the Function Call event *es* is generated by the *Chart* block. The flow control of output messages from the application to the function call block *Synchronized send* is assumed to be explicit and is not shown in this model. The send and receive activities over the network are assumed to be achieved using the xPC Target UDP/IP blocks. It may be mentioned that the input event *ea* to the chart of Figure 5.15 is programmed to activate on a rising edge generated typically by a transition of the second output of a xPC Target *UDP Receive Binary block,* irrespective of whether the implementation provides for an automatic reset.

The implementation of the other two modules, that is, the orbiter and the earth station, is identical, and the implementation of the event scheduler is left as an exercise.

The implementation of the DRTS simulating the communication between the earth station, an orbiter, and a lunar rover presented in this section assumes only a basic knowledge of MATLAB Stateflow. A more complex and better event-driven implementation has been presented by Choudhury et al. [3].

It is clear from the Stateflow charts presented in this section that the convergence of time synchronization for the DRTS is a nonzero multiple of the sampling time. Thus, a smaller sampling time produces a better convergence. Moreover, abrupt time synchronization may lead to an erroneous sequence of time stamps associated with queued messages. Thus, the *Soft Synchronization Unit* block is proposed, which is dummied in the implementation presented in Figure 5.14. Another assumption is that the event-scheduler module computes the coordinates of the orbiter, the earth station, and the rover at every sampling interval according to a prefixed algorithm. A modification of this simple scheme could be computation of coordinates from periodic messages sent by the communicating modules containing position-related information, as is done in the implementation presented by Choudhury et al. [3].

5.4 Simulating Protocols: The TrueTime Simulator

TrueTime [4] toolbox is a MATLAB/Simulink-based library of simulation blocks that extends usability of MATLAB/Simulink to simulate networked process control cosimulation of controller task execution in real-time kernels, network transmissions, and continuous plant dynamics. TrueTime blocks include generally used networks as Ethernet, CAN, TDMA, FDMA, Round Robin, and Switched Ethernet. The latest version is TrueTime 2.0, which is a Beta version but is fairly stable. The detailed instructions for installation and usage are available in Cervin, Henriksson, and Ohlin [4]. Figure 5.17 lists the different blocks in the TrueTime library that can be used with regular Simulink blocks.

The details of the blocks are available in Cervin et al. [4] along with relevant examples, and the reader is expected to be familiar with the basic usage of these blocks. There are two ways of simulating systems using TrueTime:

1. A simple case where only communication is involved, using the *TrueTime Network* or the *TrueTime Wireless Network* block along with one or more triggered *TrueTime Send* and asynchronous *TrueTime Receive* blocks.

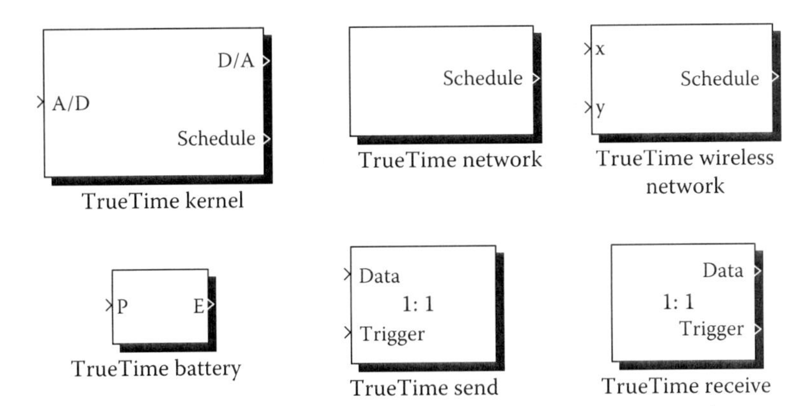

FIGURE 5.17
Blocks in TrueTime 2.0 Library.

2. A more general and flexible case where the computation and communication can be scheduled by programming the TrueTime Kernel block either using MATLAB or C/C++ functions and a set of library functions.

While the second approach is a more general approach and allows a lot of flexibility, the first approach is adequate for simple applications such as simulation of a Networked Control System (NCS) application, as shown in Figure 5.18, where a system

$$G(s) = \frac{1}{(s+1)} \tag{5.2}$$

is controlled by a discrete proportional–integral–derivative (PID) controller over a switched Ethernet connection between the sensor and the controller, and between the controller and the actuator in a control loop, as shown in Figure 5.18.

The controller chosen is a discrete PID controller and the output of the NCS on the scope *yur* is shown in Figure 5.19.

In Figure 5.19 the magenta line on the top scope represents the controller output for the system (Equation 5.2) discretized with a sampling interval of 0.001 s and exhibits discontinuity, as the loss probability in the network has been chosen to be 0.6 for the simulation in Figure 5.18, which is high.

The usage of different blocks is explained next.

5.4.1 The TrueTime Network Block

The *TrueTime Network* block configures the simulated network and is shown in Figure 5.20.

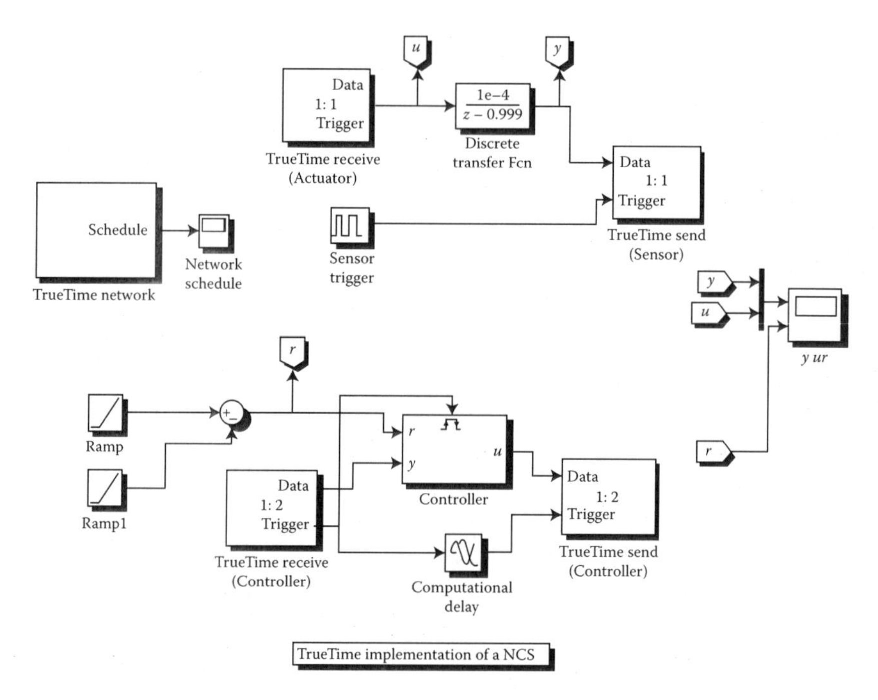

FIGURE 5.18
TrueTime simulation of a NCS.

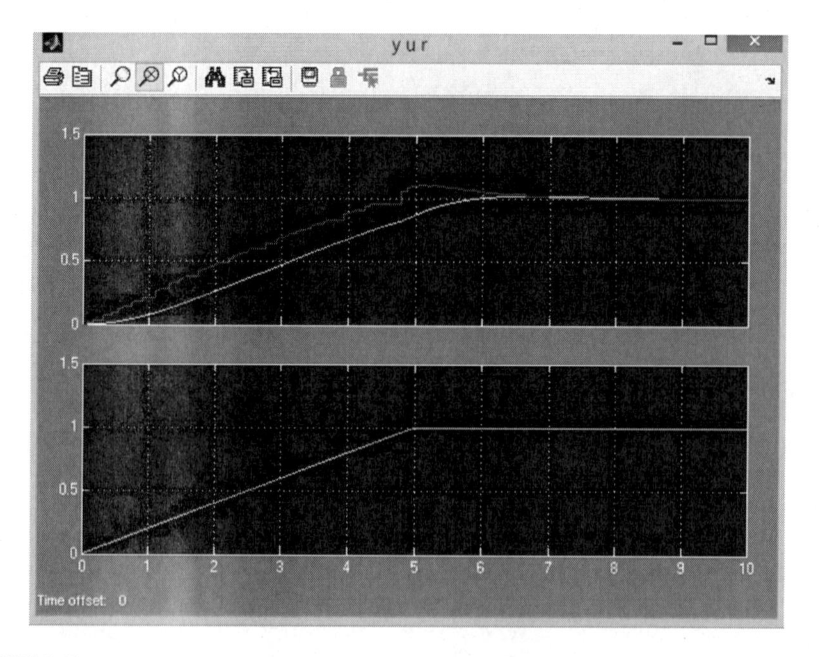

FIGURE 5.19
(**See color insert.**) Output of the NCS represented by Figure 5.18.

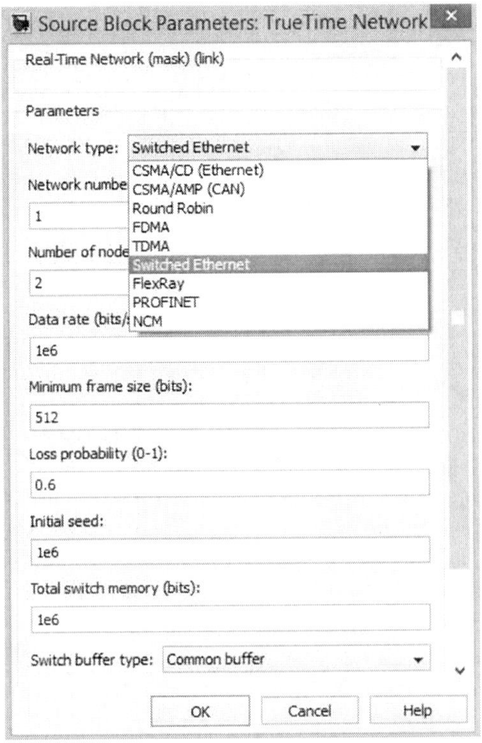

FIGURE 5.20
TrueTime Network block.

This block simulates transfer of a packet in a local network. TrueTime supports a number of protocols like CSMA/CD (e.g., Ethernet), CSMA/CA (CAN), switched Ethernet, round-robin (e.g., token bus), FDMA, TDMA (e.g., TTP), FlexRay, and Profinet. The configuration parameters are block specific and are detailed in Cervin et al. [4] and relevant parameters are replicated here for easy understanding.

Network number	The number of the network blocks in the model starting from 1.
Number of nodes	The number of nodes that are connected to a network. The nodes are identified by a tuple (network number, node number). For the model shown in Figure 5.18 this is 2. The node representing the plant is assigned node number 1 and consists of a sensor at its output and an actuator at its input.
Data rate (bits/s)	The speed of the network. This affects the propagation delay, which is usually ignored in simulation.
Minimum frame size (bits)	A message or frame shorter than this will be padded to give the minimum length. For the option chosen in Figure 5.19, it is 512 bits.
Loss probability	A real number in [0,1] that indicates what fraction of messages will be lost.

For the case shown in Figure 5.18 where switched Ethernet has been chosen, *a store and forward* is assumed with a common FIFO (first in, first out) buffer for storing messages from all nodes. The *switch memory* parameter denotes the size of this FIFO. The significance of the *schedule* output becomes evident if one examines a TDMA. The output is a vector of node IDs specifying the repetitive sequence of slots (schedule) allotted to the nodes. A 0 indicates that no node is allowed to transmit during this slot.

5.4.2 The TrueTime Send and Receive Blocks

Figure 5.21a shows *TrueTime Send* block parameters associated with the sensor node used in the simulation presented in Figure 5.18. The block is always triggered. The corresponding Receive block parameters associated with the receiver block of the controller is shown in Figure 5.21b.

The *TrueTime Send* block is always triggered. For the NCS shown in Figure 5.18, the trigger chosen is a rising edge. Thus the sensor node in Figure 5.18 is triggered at the frequency of the *Sensor Trigger*. The Receive block can be programmed to generate a trigger, for example, the TrueTime Receive block associated with the controller in the NCS shown in Figure 5.18. The controller in the NCS of Figure 5.18 is an event-triggered controller and the *Computational delay* block actually simulates the delay spent in controller computation, which then triggers a send from the controller to the actuator.

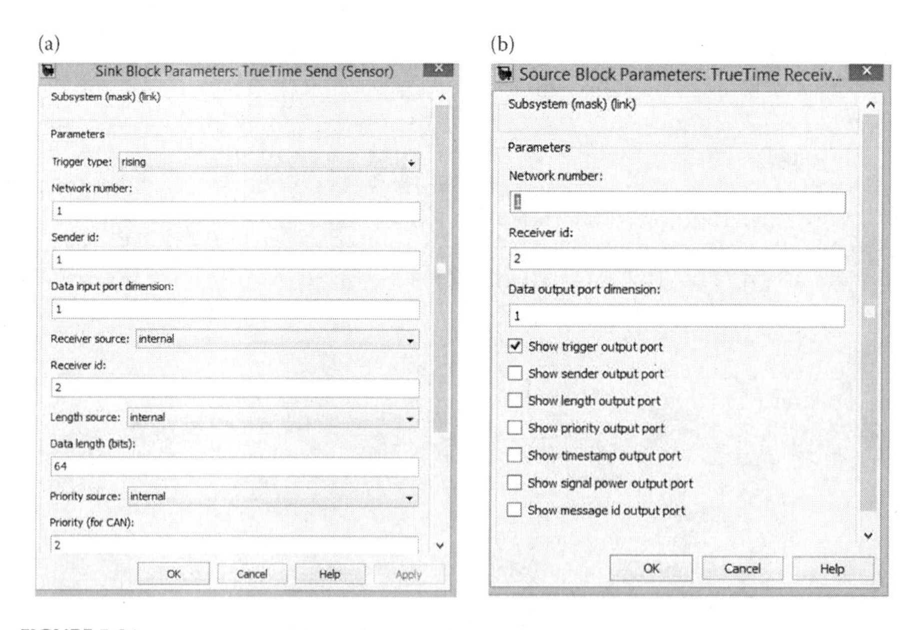

FIGURE 5.21
(a) TrueTime Send block. (b) TrueTime Receive block.

5.4.3 The TrueTime Kernel Block

The dialog box corresponding to the *TrueTime Kernel* block is shown in Figure 5.22.

The significance of the nonobvious fields are explained as follows:

Init function—Defines the name of an M-file or a MEX file where the initialization code is stored.

Battery—It is an optional parameter that can be used if device is battery-powered.

Clock drift and *clock offset*—Defines value of time difference from standard time. When the clock drift is set to 0.01 it means that the device local time will run 1% faster than real-time. Clock offset sets time offset from the first run of the simulation.

5.4.4 TrueTime Wireless Network Nodes

Figure 5.23 shows a *TrueTime Wireless Network* block, with supported protocols listed in a scroll-down.

The significance of the different parameters listed in the block are defined as follows:

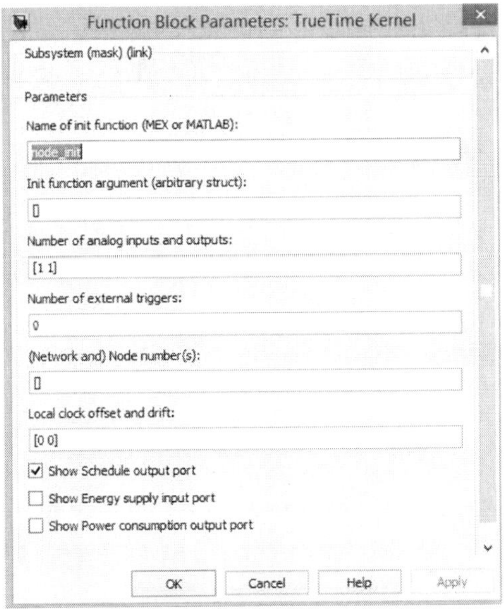

FIGURE 5.22
TrueTime Kernel block.

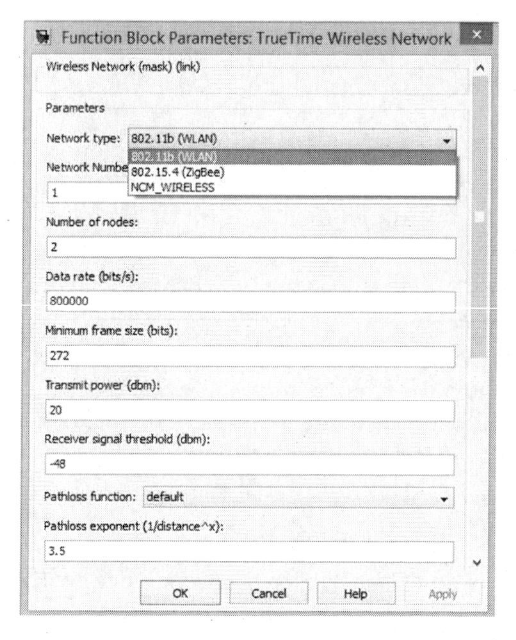

FIGURE 5.23
A TrueTime Wireless Network node.

Network Number—Starts from 1 and defines the Network block number. Wired and wireless network nodes cannot share the same number.

Number of nodes—Defines the number of nodes in the network.

Date rate—Defines the transmission speed.

Frame size—Denotes the minimum frame size in terms of bits. A message shorter than this will be padded.

Transmit power—Defines the strength of the radio signal.

Receiving threshold—Threshold value of signal above which the medium will be tagged busy.

Pathloss exponent—This is modeled as *path loss* = $(1/D^a)$. This option allows choice of the parameter a based on the medium.

ACK timeout—The timeout value for receiving acknowledgment.

Retry limit—The number of times a node attempts retransmission.

Error coding threshold—The limiting value of block error for successful message reconstruction.

Detailed documentation of the TrueTime library blocks with examples may be found in Cervin et al. [4]. A *TrueTime Battery* block is optional and can be used in a simulation to test an energy model of communication, for example, in scenarios involving wireless sensor networks.

SIMULATION EXERCISES

5.1. Develop a real-time windows target to generate a sine wave $y(t) = 2 \sin 31.4t$ and hence derive a second sine wave from it having the same amplitude and frequency and a phase lag of $2\pi/5$. Direct the outputs on a scope to generate a real-time output of both the sine waves.

5.2. Extend the Moore machine of Figure 5.12b to develop the xPC Target simulation module for the event scheduler and the full simulation of the earth–orbiter–rover communication scenario assuming that the distance between the Earth and the moon is 400,000 km and the orbiter rotates at an orbit that is 100 km from the surface of the moon.

5.3. For Exercise 5.2, if the earth station becomes the master clock, how do the simulation modules change? Develop the corresponding FSMs and the xPC Targets.

5.4. Translate the simulations of Exercises 5.2 and 5.3 to appropriate TrueTime models. In each case neglect, the transmission energy aspect and concentrate on the delay.

5.5. Wireless sensor nodes are placed radially outward from the center in a circular field such that the circular distance between two sensor nodes remains constant. Simulate a suitable traffic for the sensor network where the data from the nodes at the periphery are sent radially inward through nodes in the successive inner layers using TrueTime. Test the energy consumed using a TrueTime Battery node corresponding to the traffic model you developed.

References

1. Advantech. PCI-1171U. http://www2.advantech.com/products/1-2mlkc9/pci-1711u/mod_b8ef5337-44f0-4c36-9343-ad87d01792d1.aspx.
2. The MathWorks. xPC Target. http://in.mathworks.com/tagteam/37937_xpc_target_selecting_hardware_guide.pdf.
3. Choudhury, D., Mitra, T., Ziah, H., Angeloski, A., Buchholz, H., Landsmann, A., Luksch, P., and Gupta, A. A MATLAB based distributed real-time simulation of lander-orbiter-earth communication for lunar missions, 38th COSPAR Scientific Assembly, Bremen, 2010.
4. Cervin, A., Henriksson, D., and Ohlin, M. *TrueTime 2.0 Beta—Reference Manual*. http://www.control.lth.se/project/truetime/report-2.0-beta5.pdf.

6

Design and Testing

Computer-based implementations of real-time systems such as data acquisition systems and control systems have been in use in industrial applications for a long time. Such systems, being programmable, provide considerable flexibility, and their functionality can be modified comparatively easily. These systems are capable of being used to implement complex functions, and offer advantages of improved testing, superior user interface, and improved diagnostics support including self-checking. These systems can provide even derived information, that is, data computed based on the state of several external inputs, in real time.

Real-time systems used in industrial control are often *safety-critical systems.* A safety-critical system is defined as one in which any failure can lead to accidents that can cause damage to the system, process, personnel, and environment. Typical examples of these are control systems for control of nuclear power plants, systems used for control of spacecraft, and even in automotive control. Such systems are built around two major constituents, that is, hardware and software, and the design and implementation practices for these are quite different from each other. The flexibility of features provided by computer-based systems arises from the software constituent of such systems. However, incorporating this very flexibility to the characteristics of such systems adds complexity to implementation of the software.

Computer-based systems are vulnerable to failures like any other system; failures in computer-based systems may take place due to faults in hardware or in software. Faults in hardware may arise due to errors in design or manufacturing, and can be minimized by use of proper design techniques, exhaustive testing, and maintaining proper quality assurance during manufacturing. Hardware faults may also arise from component failures due to wear-out or degradation caused by the environmental factors and are, random in nature.

The software constituent of computer-based systems, on the other hand, differs substantially from hardware in terms of requirements for design technique, testing schemes, and reliability assessment as well as failure modes. Thus software, being a logical rather than a physical entity, is developed or engineered and not manufactured or fabricated in the sense that hardware is. Software testing can never be exhaustive enough to be carried out in cost-effective time, as it is not amenable to accelerated life testing or stress testing. Furthermore, concepts like continuity and interpolation are much more difficult to apply in the context of software programs. Software does not adhere to physical laws and does not wear out nor is it affected by environmental factors. In fact, software faults are design errors, and result from errors in requirements, design, or implementation. Thus software faults do not follow the classic bathtub curve and will normally occur at an early stage when these can be corrected. However, some design errors may remain undetected for long periods of time after the system is deployed and may occur suddenly because of some particular input from the field or when a particular execution path in the flow is traversed. Finally, software implementations may not be tolerant to small errors and the outcome of wrong inputs can be entirely out of proportion to the extent of the error in the inputs.

The use of computer-based systems in control applications provides several advantages as discussed earlier. However, to benefit from the use of such systems, these must be demonstrated to be reliable with appropriate degree of confidence. Currently such systems are not amenable to quantitative assessment of reliability primarily due to the software component of these systems. Nor can they be proved to be usable by mere testing after development is complete. This necessitates a different approach to be followed to demonstrate system reliability. It is required to address the challenge of proving software reliability by using qualitative techniques. Therefore, assessment of software has to be based on evidence that it is correct (i.e., conforms to specifications) and has a high level of *integrity* (i.e., it completely implements the requirements and is free from defects).

6.1 Safety Integrity Levels

In 1996, the Instrument Society of America enacted standard ISA S84.01, and it was subsequently revised [1] to classify safety requirements for instrumentation used in process industry. These requirements, termed Safety Integrity Levels (SILs), are measures of the safety of a given process. Each individual piece of instrumentation can be certified for use in a given SIL environment but cannot carry any SIL rating of its own.

However, the use of software in control and instrumentation systems, and the consequent introduction of systematic faults, required changes in the approach on determining the reliability of systems. Conventional methods of determining reliability are based on considering the contribution of random faults only.

Subsequently, standards emerged to help quantify safety in software-based systems. The International Electrotechnical Commission devised a standard, the IEC 61508 [2], that defines an SIL as one of four distinct levels that identify the safety integrity requirements of a system continuously performing a control function in terms of probability of failure per hour (see Table 6.1).

TABLE 6.1

SIL and Probability of Failure

SIL	Probability of Failure per Hour
4	$\geq 10^{-9}$–10^{-8}
3	$\geq 10^{-8}$–10^{-7}
2	$\geq 10^{-7}$–10^{-6}
1	$\geq 10^{-6}$–10^{-5}

The lowest level of safety integrity, SIL 1, requires following a systematic engineering process for development of a software-based control system, adherence to quality assurance techniques for development and manufacture of hardware, and development and implementation of software. This must be supplemented with an elaborate verification and validation process, proper analysis of the software, and independent assessment of the system.

Designs for higher integrity levels—SIL 2, SIL 3, and SIL 4—require increasing rigor of each of the aforementioned activities. This imposes more requirements on quality issues, for example, better system architectures, more fault-tolerant hardware designs, restrictions on software constructs to be used, and more rigorous review process.

Each industry (automotive, defense, nuclear, etc.) defines SILs in its own context, but the essence is that with increasing requirements of reliability of the systems, increasing rigor has to be followed in the design, development, and review of the systems. Building in quality in software presents the most significant challenge.

As stated earlier, it is not possible to demonstrate absence of faults in software and it is impractical to ensure reliability of software merely by exhaustive testing. It is possible to deal with only qualitative measures of software reliability. Thus the higher the reliability requirements of a software-based system, the greater the rigor that must be used in following development practices, analysis testing, and reviews.

Correctness and integrity of software must be ensured by developing a systematic, carefully controlled, and fully documented process, which is subjected to exhaustive third-party review activity. The software used in nontrivial applications is generally complex and, therefore, its quality must be ensured by monitoring it at every stage of development. This requires that a structured life cycle approach be followed. Since the hardware and the software in a computer-based control system work in tandem, a structured life cycle approach should indeed be followed for the design and implementation of the complete system. This is elaborated upon in the next section. Precise and detailed documentation must be produced by "designers," and this documentation must be concurrently reviewed by people other than "designers" to provide assurance that good design practices have been followed and sufficient analysis and testing have been performed.

6.2 System Development Life Cycle

The life cycle of computer-based systems consists of the entire stretch from defining the requirements through the development and installation and commissioning, to the operation of the system (Figure 6.1). The prerequisite for entering the life cycle is the existence of a statement of purpose that defines the role of the computer-based system; thus the system requirements are the driving force behind the entire development life cycle.

During the *requirements definition* phase, the design engineer must generate the system requirements, which describe the key requirements of the system at a black-box level. This includes the role of the system, various modes of its operation, and functional and performance requirements in each mode; its external interfaces; environmental constraints; and miscellaneous aspects including, maintainability, safety, and security. Thus the total behavior of the system is defined based on the overall user requirements without any statement on the means of its implementation.

This activity is very important because any errors in the definition of requirements will ripple through the entire design process and may get detected only when the system is deployed in the actual environment. In some cases, requirements are not fully thought through and when undefined or unanticipated scenarios occur, the system fails to produce the expected outputs. The scenarios that have not been stated in the requirements may arise because of unanticipated failure conditions or even implicit but unstated normal conditions.

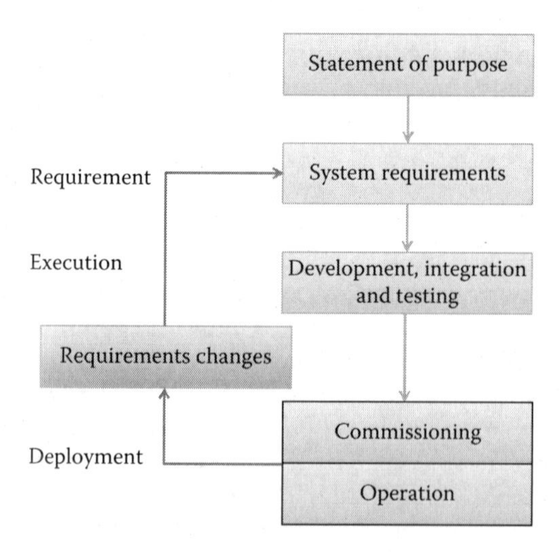

FIGURE 6.1
System development life cycle.

These kinds of errors have to be prevented at the requirements definition stage by thorough evaluation of the environment that would prevail during deployment of the system. It is required to ensure that everything that is applicable in the context of the relevant system is identified and stated explicitly. It is also prudent to have an exhaustive dialogue between the developer and the user prior to commencement of development to eliminate any misinterpretation of the stated requirements.

Some scope for misinterpretation of requirements arises because of the ambiguities associated with the usage of natural language, such as English. This scope for misinterpretation may be minimized by usage of semiformal (e.g., UML) or formal language (e.g., Z) to state the requirements. It is frequently observed that the effort required to correct a requirements error is far less if done at the stage of defining the requirements. Any error that is overlooked during the requirements generation phase is far more expensive to correct if it is done during the testing phase and still more so if the error correction is done after deployment of the system.

A couple of examples of well-known dangerous situations having arisen because of unstated requirement conditions are provided in Figure 6.2.

The entire development cycle is carried out during the *execution* phase and includes a detailed design of the system and its constituent hardware and software, implementation of the physical system, and testing of the same to ensure conformance to requirements enunciated earlier. The detailed design is marked by definition of intermediate milestones and generation of design documentation. Standardization of the complete development process ensures that (repeatable) quality can be achieved. This requires, for example, systematic planning to define the technical and managerial project

Examples of Requirements Errors

During the Cold War, both the USA and USSR were geared to detect missiles launched by the other country and to respond quickly and strongly in accordance with a mutually assured destruction doctrine.

- In 1960s, the US Ballistic Missile Early Warning System generated a report of a missile attack. It turned out that radar in Greenland had caught echoes from the moon. The possibility of receiving such echoes had not been anticipated during the requirements definition.
- Likewise, in 1983, the Soviet early warning satellite system picked up the sun's reflections from the top of the clouds and interpreted these as missile launches. Again, the possibility of receiving signals from such reflections had not been anticipated during the requirements definition.

Fortunately in both events, the errors were identified in time and a global nuclear war was averted.

FIGURE 6.2
Examples of errors arising out of limitations in specifying requirements.

functions, and devising a plan to maintain consistency of various hardware and software items at all times.

As stated earlier, such systems must be demonstrated to be reliable with an appropriate degree of confidence. It is necessary to devise plans to define the scheme to carry out verification activities on the entire system as well as hardware and software at various stages of the development process. Finally, it is necessary to devise plans to carry out quality assurance at relevant stages of hardware production and software development.

The *deployment* phase consists of the installation of the system at the final site of application and connecting all field inputs and outputs to the system. The system is tested to ensure it behaves as expected under all conditions for which it has been designed. The system is then treated as commissioned and its behavior during operation must be tracked. Should any undesired behavior occur, or, any unforeseen situations arise, the original requirements need to be modified accordingly. Such change in requirements leads us back to the execution phase and these revised requirements have to be incorporated in the design. The structured development process followed earlier has to be carried through again.

Our focus in this chapter will be on the execution phase and details of the structured process carried out therein.

6.2.1 Verification and Validation

The process of verification and validation (V&V) is a set of review activities performed concurrently on the entire system as well as on the hardware and the software throughout the system development life cycle to ensure that the system meets the specified functional, performance, and reliability requirements. It also provides assurance that faults that could degrade system performance have been identified and removed before the system is deployed for use.

Computer-based systems are being used to implement more complex functions, and the possibility that designers will make errors has to be recognized. These errors can result in undetected faults in the system that could in turn result in dangerous failures. Hardware development processes are generally considered to be sufficiently mature to effectively detect systematic errors. However, software development is a complex process and faults can be introduced at every stage of the development process. Hence its quality must be ensured by monitoring it at every stage of a structured software development process rather than by merely testing it after the development is completed. This makes it necessary to apply active measures in a systematic manner to find and correct errors. This places a heavy dependence on V&V to detect errors and successful V&V will result in the reduction of the number of software faults to a tolerable minimum.

The system development life cycle is structured as a set of stages, each of which leads to a deliverable product, usually documentation. Verification is an intensive process carried out at the end of each stage of the life cycle to ensure that the delivered product conforms to all the requirements, thereby ensuring that errors are uncovered and corrected as early as possible in the life cycle. Verification includes a detailed review of documents, review of software code, and testing (of hardware or software) at each stage to confirm readiness for the development process to move to the next stage. The process of verification may be carried out by confirmation of facts by direct examination or by review of objective evidence that specified requirements have been fulfilled. The development life cycle itself is commenced after the initial verification process determines that the requirements for a system are complete and correct.

The validation process is carried out at the end of the development cycle on the final implemented system to confirm that the hardware and the software function together in accordance with the stated system requirements. System validation involves evaluating the external behavior of the system and includes some testing in an environment that is representative of what would be experienced by the deployed system. Successful completion of the validation process provides assurance that the complete system (including its hardware and software constituents) functions such that it meets all specified requirements. However, the final element of the development life cycle is validation of relevant features of the system after it is deployed in the end-use environment.

In general, the team of engineers that perform V&V activities should be independent of the team that carries out the development activities for the system under review. The V&V team may be composed of engineers who have experience in development of other comparable systems but are not directly involved in the development effort of this system. The expertise of the V&V team should be comparable to that of the development team. The team may directly perform verification reviews and may also review results of analysis carried out by the development team. It should provide its review and verification reports to the development team to help improve the quality of the design.

6.2.2 Detailed Execution Cycle

The first step in the development process is to develop the system architecture, including partitioning of the system into multiple subsystems if necessary. Depending upon the size of the system and the complexity of requirements, the system is progressively decomposed into various functionally independent subsystems, which in turn are further decomposed into hardware modules and software packages. Each subsystem is allotted functions (from among the total set of requirements) that can be performed largely independent of the functions of the other subsystems. This helps each subsystem function efficiently with the least dependence on communication of information from other subsystems. Simple systems may not require any

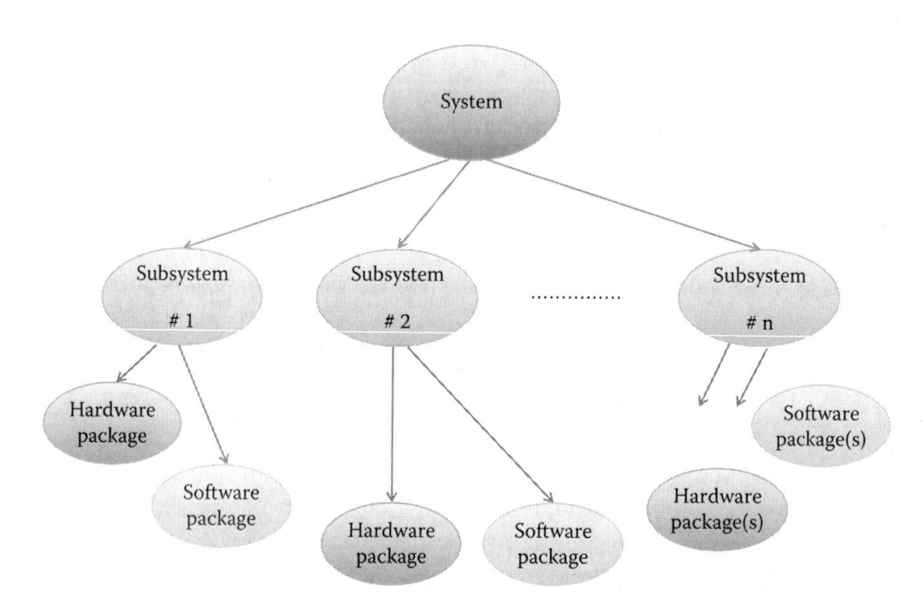

FIGURE 6.3
System decomposition.

such decomposition of structure and functionality. Finally, all subsystems put together must perform all the requirements that are intended for the complete system. The decomposition is iterated as shown in Figure 6.3 until a stage is reached when the development of these blocks (hardware and software) can be done separately and the same can be implemented with view to eventual integration. Hardware modules and software packages common to multiple subsystems can be developed independently. A procedure must be devised to provide the scheme to carry out integration of the complete system from its constituent subsystems and for testing of the system.

As a first step, the system architecture should be verified against the system requirements to ensure integrity and completeness, that is, all stated requirements have been carried over into the first stage of the design. All inputs from the process being controlled have been allocated to one or more subsystems and likewise all outputs to be generated by the composite system are allocated to one subsystem or another. The interaction between various subsystems must also be defined so that data and information required by one subsystem from other subsystems is clearly identified at both ends. The subsystem that has to send information to another subsystem must list this activity as a requirement to be met by it. Likewise the recipient subsystem must also list the need to obtain such information from a peer system as a requirement.

Each subsystem is required to be developed largely independently until the design and implementation and testing are completed. The entire process has to be carried out systematically and is divided into two main activities: the *development process* and the *implementation process*.

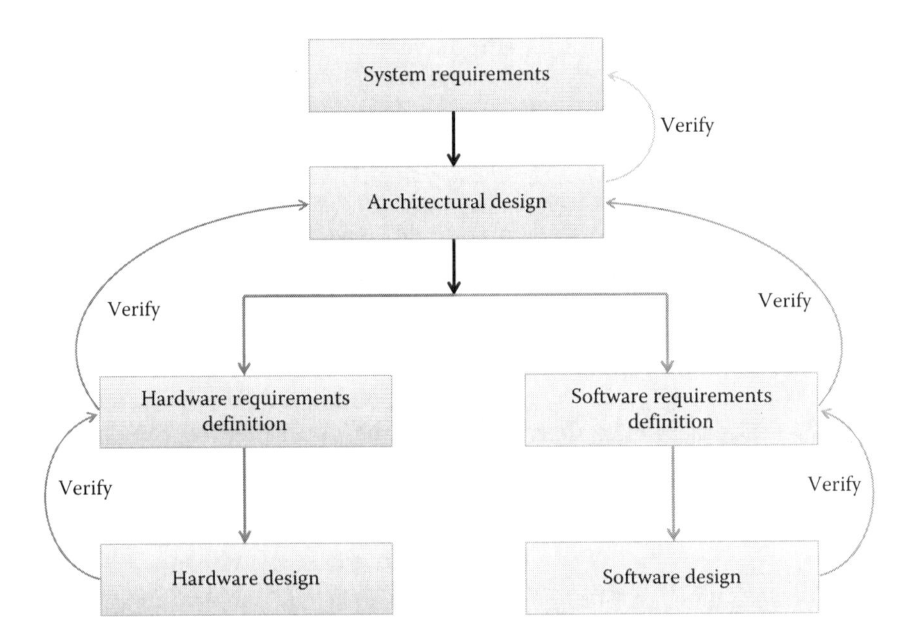

FIGURE 6.4
Development process.

During the detailed development process, as shown in Figure 6.4, the architecture of the subsystem has to be defined. The total functionality to be carried out by the subsystem has to be suitably allocated for implementation to hardware and software. The development of the constituent hardware and software portions of the subsystem can be carried out as parallel and nearly independent processes. The design process results in production of several documents; each of these documents may have to be produced for each of the subsystems defined during system decomposition. Each of these documents should be verified against the upstream document to ensure integrity and completeness. The decomposed structures have to be independently developed and tested, and then, as an activity reverse of the decomposition process, must be progressively integrated and tested. Documented procedures must be defined for the entire process of integration and testing at each level of hardware and software as well as each constituent subsystem.

It is necessary to generate the hardware requirements document for each subsystem to provide details of the functions to be performed by the hardware package. Several other aspects of the requirements that need to be identified include interfacing with the field inputs, interfacing with other subsystems, testability requirements, environmental requirements, and power supply requirements. It should also list the applicable constraints/standards. The definition of the hardware requirements should be unambiguous, complete, and consistent. Again, the hardware requirements for each subsystem should be verified with the system architecture document to ensure all relevant requirements have been carried through correctly throughout the complete cycle.

The next step is to design the hardware for each subsystem by partitioning it into modules, if necessary, carrying out the detailed design of each such hardware module, and providing design details and complete inputs for the manufacturing process. The complete information should be used to generate a hardware design document for each subsystem. This document should be verified with the hardware requirements document for the corresponding subsystem to ensure all relevant requirements have been carried through correctly.

The software requirements document identifies the subset of the system requirements for each subsystem that is to be implemented using software. It must provide details of the functions to be performed by the software, its performance and testability requirements, applicable constraints, and so forth. The definition of the software requirements should be unambiguous, complete, and consistent. The software requirements for each subsystem should be verified with the system architecture document to ensure all relevant requirements to be implemented in software have been carried through correctly.

The software design activity comprises two steps, the first being to design the software architecture that is, separating the software package into functionally independent modules that can be developed in parallel and defining the function of each as well as the interfaces between various modules. The second step is to carry out the detailed software design that is, the actual implementation of these modules including defining data structures, algorithms, and the external interfaces. The complete information should be used to generate software design documents for each subsystem. These documents should be verified with the software requirements documents for the corresponding subsystem to ensure all relevant requirements have been carried through correctly.

Consequent upon completion of the development process, the hardware and software have to be fabricated/developed and tested during the implementation process as shown in Figure 6.5.

Each of the identified hardware modules has to be fabricated and tested and the results documented. This document has to be verified for each individual module against the hardware design document to ensure that the design requirements have been met. The hardware modules have to be assembled to achieve each individual subsystem and these subsystems have to be tested (refer to Section 6.5.1) in accordance with the requirements. The results must be documented and verified against the corresponding hardware design documents.

Likewise, the software for each subsystem has to be coded, tested at unit level, and integrated and tested (refer to Section 6.4). The resultant documentation has to be verified against the software design document to ensure that the results are in accordance with the design.

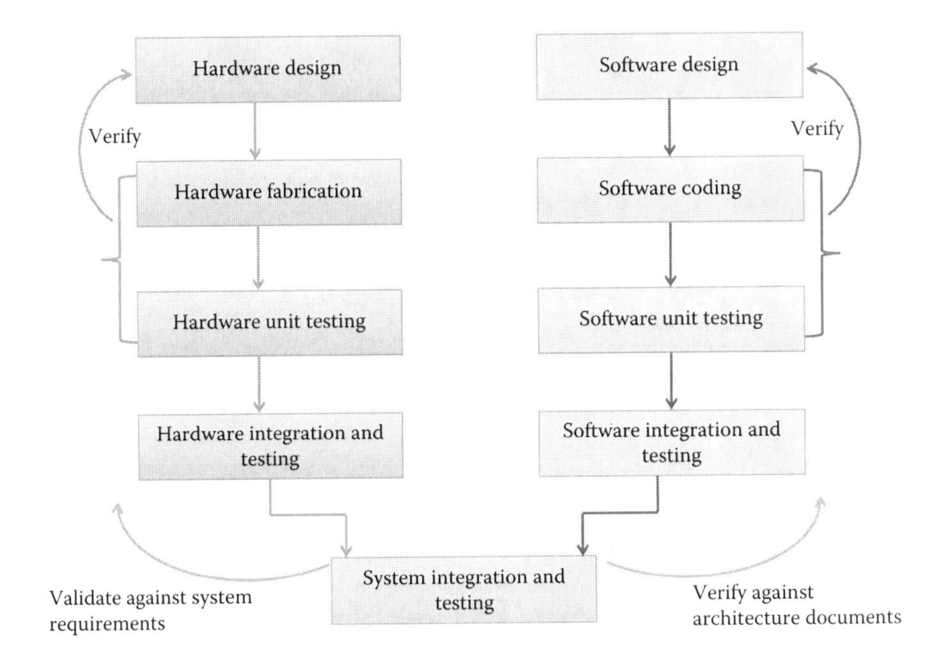

FIGURE 6.5
Implementation process.

Each subsystem should be integrated and tested (see Sections 6.5.2 and 6.5.3), and the resultant documentation verified and tested against the system architecture design documentation.

Finally, the complete system has to be validated (refer to Section 6.5.4), that is, checked to ensure that it functions in accordance with the system requirements.

Having defined the system development life cycle and identified the steps involved in the complete process, it is worthwhile to view the design considerations to be invoked during the process of the detailed design at each level.

6.3 System Design

A distributed real-time system used in a safety-critical system such as a control system must be dependable, that is, it should provide high availability (be in service for long periods) and high reliability (perform correctly as per design). This dependability may be achieved by minimizing the possibility of occurrence of faults, tolerance of faults if these do occur, and by correction of faults. Faults may be prevented by following good design practices (following recommended standards, keeping sufficient design margins, etc.)

and manufacturing/development practices for hardware and for software respectively (practicing stringent quality controls, etc.) Faults may be tolerated by building in sufficient redundancy (and diversity for more critical applications) into the system hierarchy to allow the system to continue to function properly or, in degraded mode, even in the face of some failures. This makes it necessary that partial failures are detected and repaired early by building in exhaustive system diagnostics and taking corrective action when called for. Faults may be corrected by replacement of the faulty components in hardware and bug correction in software.

Computer-based control systems have four major components that are to be addressed during the design phase:

- Architecture
- Hardware
- Software
- Human–computer interface

6.3.1 System Architecture

The system architecture is defined by carrying out the logical structuring of the system, decomposing the complete system into a hierarchy of subsystems, if necessary, defining these subsystems and the interfaces of these subsystems with each other and with the external world. The functionality of each subsystem and of hardware and software in each subsystem is defined such that the complete desired functionality is achieved and all performance requirements are met. The design requirements applicable to hardware and software are defined more elaborately to serve as the basis for more detailed design and subsequent testing of the composite system.

Several design principles are to be followed in this process of decomposition to ensure reliability and ease of verification. Standard concepts like segregation of critical functions from noncritical functions, *defense in depth, single failure criteria, common cause failure, fault tolerance, redundancy, diversity* are applicable here.

Several types of architectures may be selected for use in a computer-based control system depending upon the requirements of availability and reliability for the application, that is, based on the SIL of the process environment in which the system must function. The major emphasis in the choice of architectures is to determine the degree of redundancy requirements. For more critical applications, a greater amount of redundancy of the various building blocks is required. Thus redundancy can extend to processor hardware, peripheral hardware, networks, user interface stations, field signals, controllers, and so on. The specific architecture to be chosen for a given application is determined by the risk assessment carried out and the degree of safety level required to be achieved.

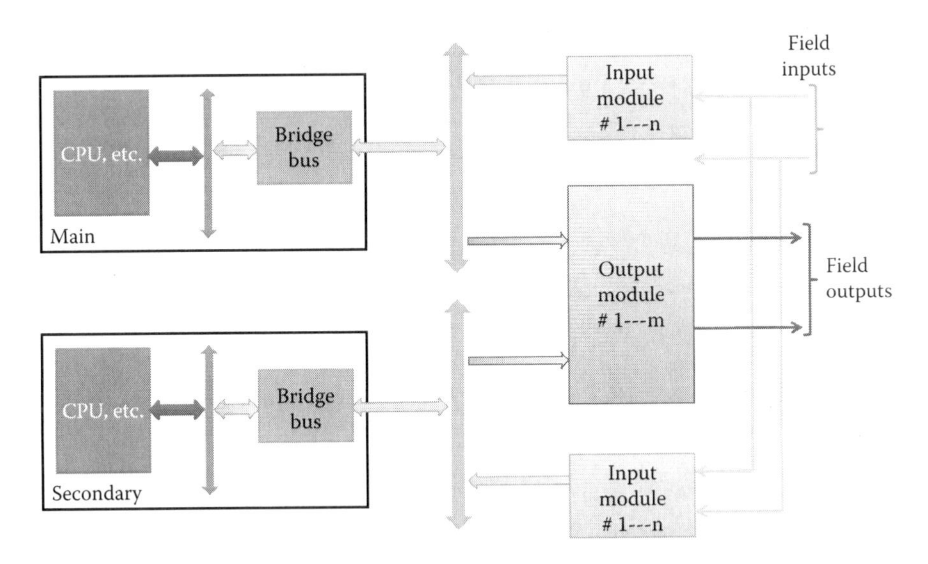

FIGURE 6.6
Typical hot standby configuration (schematic).

Two simple schematic architectures are provided here as examples. Figure 6.6 shows a simple scheme for implementing fault tolerance using two computers: a main computer and a secondary computer operating in hot standby mode. In this configuration, each computer can simultaneously and independently obtain the state of the inputs from the process being controlled and can carry out the required computation and generate the control outputs. However, the control outputs generated only by the main computer are passed through to the process. If the main computer is detected to have failed, the control outputs generated by it are blocked and those generated by the secondary computer are passed through to the process.

Thus the scheme allows the control function to be carried out even in the event of failure of one computer and the switchover from one to the other takes place in a "bounce-free" manner. The failure of the main computer is detected by a watchdog circuit (not indicated in the figure), which also controls the selection of the outputs from the computer in operation. This architecture mainly guards against failure of one computer system or gross failure of one input system but not against failure of the output system.

Figure 6.7 shows a scheme for implementing fault tolerance using a triple modular redundant configuration. This configuration consists of three independent electronic channels: Ch-A, Ch-B, Ch-C. The use of this configuration in highly critical applications demands that each parameter being monitored must have triplicate sensors. Each electronic channel receives corresponding signal inputs, that is, for each parameter it receives a signal from one of the three sensors, the first channel from the first sensor of each parameter,

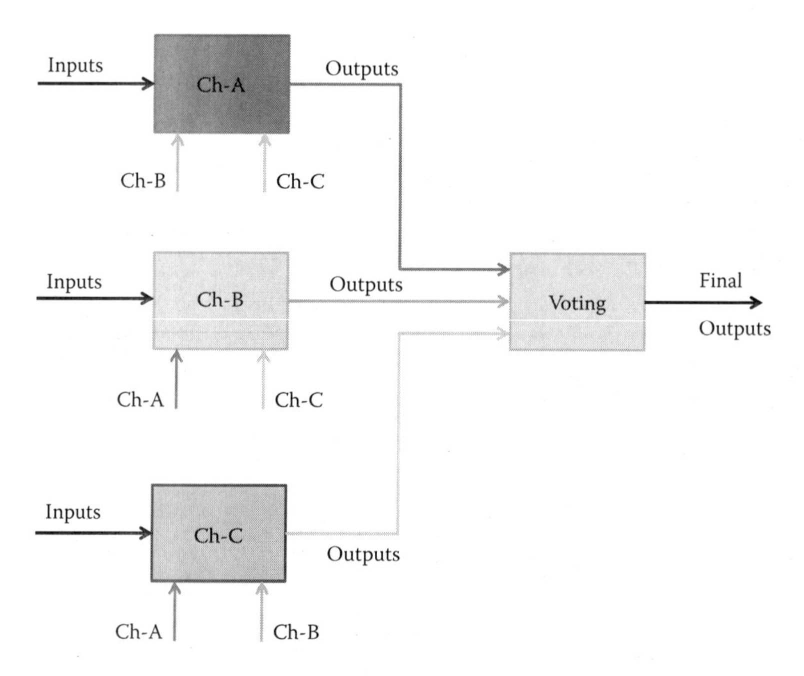

FIGURE 6.7
Triple modular redundant configuration (schematic).

the second channel from the second sensor of each parameter, and so on. Each channel processes its corresponding set of signals, implements the required control algorithms, and generates its outputs. The three channels may exchange data with each other using isolated, redundant communication paths to carry out functions like internal state comparison, channel disagreement, and channel synchronization. The signal travel paths as well as the channel electronics must be galvanically isolated from each other.

The outputs from each channel are voted upon to produce the final output; voting is two out of three ladders for digital outputs and median for analog outputs. In order to achieve still higher availability, each channel may be designed using dual redundant hardware.

6.3.2 Hardware

The design of hardware is carried out after identification of all functions to be carried out in hardware has been completed. The development process for hardware comprises partitioning the hardware modules to achieve the functions of providing computing and storage resources, data acquisition, data dispatch, and, in systems with multiple nodes, facilities for inter-subsystem communication.

Thus the hardware for computer-based systems is usually built around a bus-based architecture and consists of a processor module, I/O modules, communication modules, signal conditioning modules, and so on. The hardware modules must meet the desired functional and performance requirements, and incorporate typical features like galvanic isolation, diagnostics capability, fault tolerance, and failsafe output generation. Each of these boards is designed (perhaps based on use of hardware description languages) using a set of electronics components. The hardware implementation process consists of translating each function into actual circuits to realize each of the functional boards. The total hardware required to implement a control system is configured by utilizing a desired subset of the electronics boards available.

The subject of quality assurance for hardware components is a very well-established field, and the prevalent standards on component selection, and board design, fabrication, and testing can be followed.

6.3.3 Software

The design of software is carried out after identification of all functions to be carried out in software has been completed. The development process for software comprises decomposing the process-level functions into smaller manageable units and designing real-time processes and their execution control flow (i.e., fixing their relative priorities and their mutual synchronization), designing appropriate shared and local data structures, and interrupt and exception handling, and, in systems with multiple nodes, the internodal information transfer scheme.

The implementation of software consists of translating each of the modules into source code using the language selected for coding. It is essential that errors be minimized by avoiding complex and tricky designs, incorporating error-detection features, incorporating testability, and by keeping safe and adequate margins on critical design parameters. The code thus obtained is then compiled, linked, executed, and debugged to correct all translation errors. Limited testing is performed to ensure correct implementation of algorithms and logic.

The evidence of proving software reliability has to be provided by demonstration of the correctness and integrity of software. The main contributors to software errors are incorrect or incomplete statement of requirements, and, software design deficiencies. The issue related to incomplete or incorrect definition of requirements was addressed earlier (Section 6.2).

The second contributor to errors is software design deficiencies, which arise from the specific implementation scheme for the software. Among the most prevalent design errors are buffer overflow, access a buffer with incorrect length value, array bound violation, improper check for exceptional conditions, allocation of resources without limits, and race condition.

These kinds of software design errors have to be minimized by following recommended good practices. Coding standards (e.g., IEC 60880) are available for each of the major sensitive and vulnerable industries, and compliance with the recommendations of these can result in production of "good"

Examples of Design Errors

- Ariane 5, Europe's satellite launching rocket, reused software code from its predecessor Ariane 4, including code to convert a 64-bit floating point number into 16-bit signed integer. During its launch on June 4, 1996, the Ariane 5's faster engines resulted in larger than Ariane 4 numbers and triggered an overflow condition in this code leading to crash of the primary as well as the backup computers. The failure of both the computers activated the self destruct mechanism and the rocket as well as its $500 million satellite payload disintegrated 40 s after launch.
- The Therac-25 was a radiation therapy device that could deliver two different kinds of radiation therapy: either a low-power electron beam (beta particles) or X-rays. The latter were generated by smashing high-power electrons into a metal target positioned between the electron gun and the patient. In this "improved" version of the machine, the physical safety interlocks to position the metal barrier were replaced with software control. A bug in the software, an "arithmetic overflow," sometimes occurred during automatic safety checks and led to high-power beams being fired directly into the patient without the physical barrier coming into position. During 1985–1987 at least five patients died because of radiation overdose and several more were critically injured.
- During the war in 1991, the US deployed the Patriot Missile system to intercept Iraqi SCUD missiles. The tracking software for the system used the current time to predict the path of the target missile. A known bug in the targeting software—"drift"—in the internal clock would be fixed by rebooting the system periodically thereby resetting the internal clock. On February 25, 1991, when an Iraqi missile was detected by the system, its internal clock had drifted by one-third of a second because of not having been rebooted for about 100 hours so its computed location was predicted to an error of one-half kilometer. Not finding the missile where it was predicted to be, the system cancelled the interception and the missile went on to hit a US airfield in Saudi Arabia killing 28 and injuring about a hundred soldiers.
- The software design of the navigation system of the Mars Climate Orbiter used imperial units of measurement instead of the metric system specified by NASA. In 1998, the $125 million spacecraft made errors in attempting to stabilize its orbit and crashed into the Mars planet.
- In August 2003, about 55 million people in Northeast US and Canada were left without power, many for several days; a software bug prevented the blackout from being averted and 256 power plants went offline.

FIGURE 6.8
Design errors.

software. Several constraints have been proposed on using programming languages by, for example, MISRA C for C language, to reduce scope for misuse of some language constructs in sensitive applications.

Several examples of well-known dangerous situations that lead to loss of life as well as loss of expensive equipment because of software design errors are listed in Figure 6.8.

Quality has to be built into software during the entire development process and this includes the use of proper methods of engineering. Several development models for construction of software are pursued in industry, for example, the *waterfall model* as shown in Figure 6.9.

The waterfall model of development is a highly structured scheme in which each phase of software activity must be completed before the next phase is commenced. It recognizes the need for feedback loops between steps in contrast to the earliest simple model of code and fix.

The waterfall model is the earliest structured but simple scheme followed and works well for small projects. Changes in requirements or correction of errors detected at late stages can prove quite time-consuming and expensive to implement for any reasonable-sized projects.

In the *iterative model*, as shown in Figure 6.10, the complete software is developed in a series of iterations. The process is carried out by breaking up the software of a large application into a number of portions. The iterative process of developing and testing software is commenced by implementing

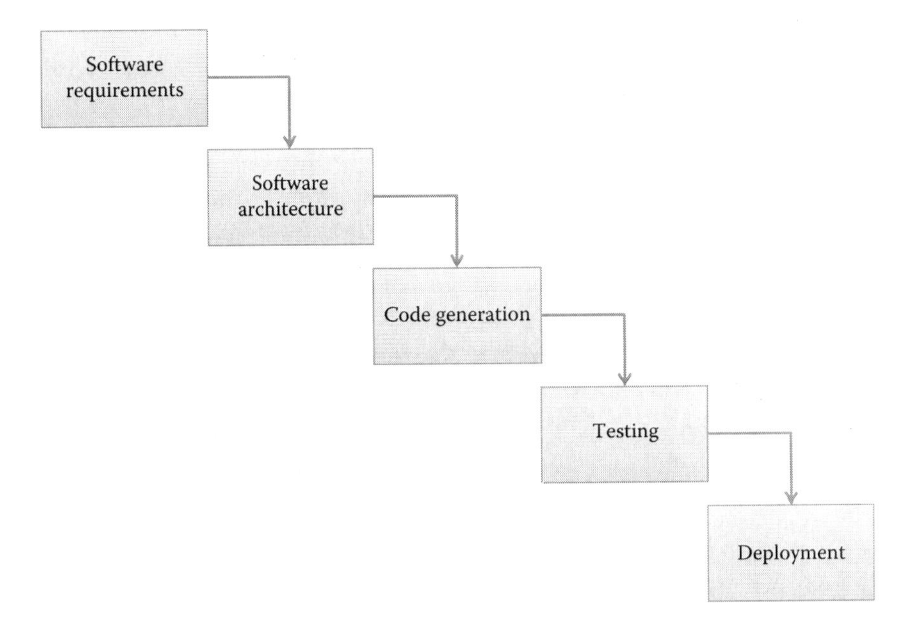

FIGURE 6.9
Waterfall model for software development.

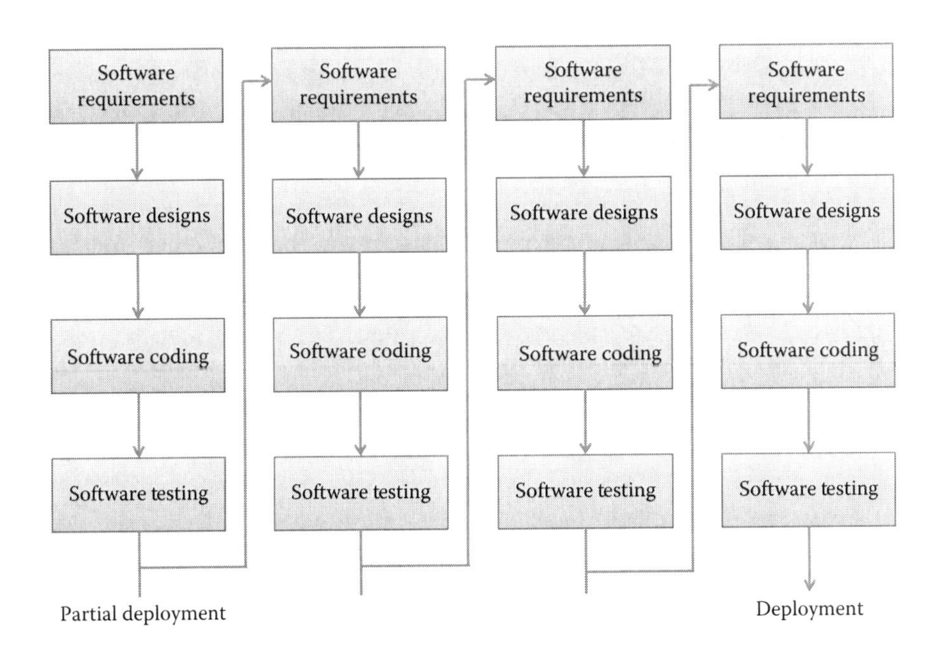

FIGURE 6.10
The iterative model.

a logical subset of the total requirements. In the next iteration, this piece of software is enhanced to implement an additional portion of the requirements. On completion of development and testing of this iteration, the process is repeated until the entire set of requirements has been implemented. Some amount of parallelism in development of software corresponding to different subsets of the requirements can be practical in this model.

The iterative model is often used in conjunction with the *incremental development* model. In this software development, work is commenced on what may be regarded as the core portion of the application. The activity for the identified chunk of the software proceeds as in the waterfall model, that is, requirements analysis, architecture design, code generation, and testing. This portion of the software that has been completed may be delivered for initial deployment in the field. This enables more thorough testing in field conditions, and if the chunk is carefully identified, it will provide service until additional iterations/increments are carried out and delivered.

The iterative model allows for some set of requirements to stay flexible for some portion of the development cycle. This can have a negative impact if the set of requirements defined at a later stage impacts the development already carried out. The management of the development cycle can also prove to be quite complex for very large projects.

The *spiral* model, as shown in Figure 6.11, combines the concepts of the iterative model with the structured scheme of the waterfall model. The model

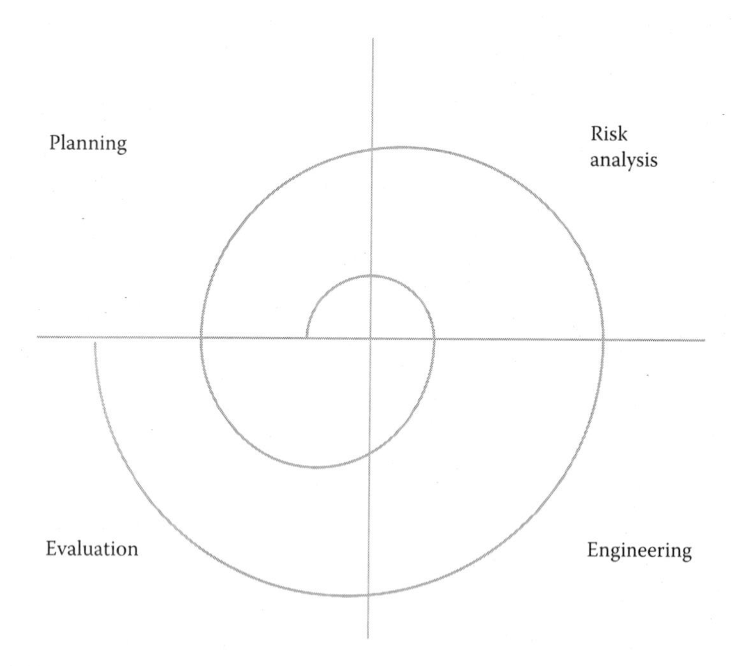

FIGURE 6.11
The spiral model.

consists of four recurring phases: planning (establishing requirements), risk analysis (identifying constraints and defining alternatives), engineering (development and testing), and evaluation (testing for suitability for eventual deployment). The development life cycle progresses in several spirals through each of these phases.

The spiral model makes it easier to accommodate changing requirements, and development can be divided into smaller parts. The management of the project is more complex and the scheme inherently demands much more documentation because of the number of stages involved.

6.3.3.1 Model-Based Design

Model-based architecture and design methodology is increasingly being used for software for critical control applications. Model-based design of software supports requirements, design, coding, analysis, and verification activities, beginning with the conceptual design and continuing through the entire life cycle, as shown in Figure 6.12. Development tools available support building models for complex functional requirements leading to rapid prototyping, software testing, and verification. Models are executable so they can be used for validation; formal verification can also be carried out on the same model. Code can be generated from models automatically using a certified code generator.

The model-based design methodology is significantly different from the conventional software design methodology. Instead of writing extensive

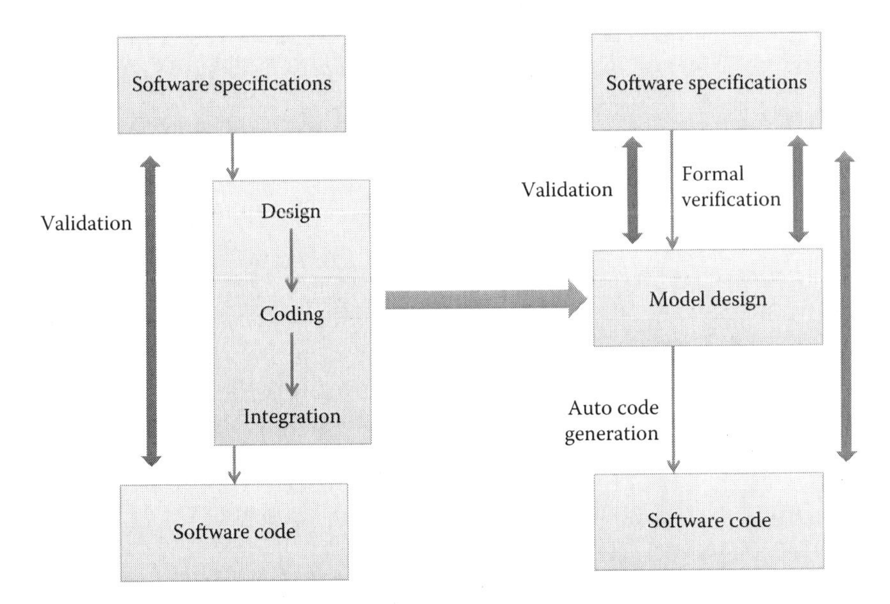

FIGURE 6.12
Model-based development of software.

software code, designers can now represent complex functional requirements using building blocks defined in terms of primitive functions having precise meanings. Models present the full system behavior in an easy-to-understand manner with use of domain-specific notations (e.g., control flow diagrams, state transition diagrams) without getting immersed in a lot of design details.

Some of the characteristics that are common to industrial control systems as well as human–computer interfaces include parallelism, determinism, temporal requirements, and reliability.

Given the parallelism between the control system and its environment, it is convenient and natural to design such systems as sets of parallel components that cooperate to achieve the intended behavior. These systems exhibit *determinism*, that is, they always react in the same way to the same inputs. The temporal requirements are determined by the process dynamics and the system must respond to stimuli from the process in desired time. The reliability of these systems is very important, as has been discussed earlier, and this determines the rigor of the design and verification.

Synchronous languages [3], which were introduced in the 1980s, allow specification of parallel and deterministic behaviors, and allow formal behavioral and temporal verification. These include *Esterel* [4] and *Lustre* [5] and the advantages of these include the availability of idealized primitives for concurrency, communication and preemption, rigorous semantics, and a powerful programming environment with the capability of formal verification.

Esterel is an imperative synchronous programming language that, besides variables, manipulates signals and is suited for describing controls. A program comprises a set of concurrently running threads whose execution is synchronized to a single clock. A signal can be an input signal (its presence can be tested), an output signal (it can be emitted), or a local signal (it can be emitted and its presence can be tested). The communication mechanism between threads is the synchronous broadcast, that is, any signal emitted by someone at a given instant is received by everybody at the same instant.

There is no notion of physical time inside a synchronous program, but rather an order relationship between events (simultaneity and precedence). The physical time is thus an external signal, like any other external signal. Each component of the parallel construct can react independently to its signal.

Statecharts are hierarchical state machines with a graphical representation involving states and transitions. There are three types of states and the machine can transit from a source state to a target state, triggered by an event, under a particular condition and a defined action is taken. A statechart incorporates the idea of a history state, which keeps track of the substrate most recently visited. A typical example is MATLAB Stateflow [6], which was covered in Chapter 4.

Lustre is a declarative synchronous language that supports writing dataflow as a sampled stream of data. Each variable is a function of time and is either an input or is expressed in terms of other inputs. A program in Lustre defines a function from its input (sequences) to its output (sequences), and output flows are defined by means of mathematical equations. Data traveling along the "wires" of an operator network can be complex, structured information. This allows the rigorous mathematical semantics of checking types of variables and their allowed compositions.

The *SCADE* modeling language [7] supports a combination of dataflow equations, for example, Lustre and safe state machine (Esterel and restricted statecharts), for modeling a system. A dataflow equation describes the logical relationship between input and output variables. Safe state machines describe the control flow in the system in terms of state transitions. The SCADE model is graphically represented as network of operators transforming flows of data.

The SCADE Suite is an integrated design environment that covers requirements management, model-based design, simulation, verification, and certified code generation. SCADE has been used in many critical industrial systems; these include the protection system of the Schneider Electric N4 French nuclear power plants, fly-by-wire controls, and automatic flight controls of the Airbus A-340.

6.3.4 Human–Computer Interface

The design of the human–computer interface, namely user interface, has a tremendous impact upon the usability of a system. The design must take into account human factors engineering and ergonomic issues to make it user friendly. This includes details of user interfaces, user functions required at each interface, sequence of dialogue between user and computer, consistency, and help features. The commands and functions, usually menu based, must be well structured in a hierarchical manner, should be grouped together in logical sets, and should be unambiguous.

The data presentation should preferably utilize mimic diagrams, which are representative of the process being controlled, and the data should be presented in appropriate graphic and numeric formats. The user should be able to view current, historical, and archival information about the process parameters. Information that is important should be presented in a prioritized manner. The system should also guard against user data-entry errors by checking for inconsistency in data entry. The interface should allow for easy navigation to enable a user to quickly reach desired information.

Finally, the human–computer interface (often called *graphic–user interface*) features have to be implemented, primarily in software, although there could be some impact in hardware, for example, for building in security features.

When the definition of the system architecture and human–computer interaction has been completed, and the allocation of functions to hardware and software defined, the design of hardware and of software can proceed independently to intersect again only when both are ready for integrated testing.

On completion of the design, development, and production cycle of the hardware, software, and human–computer interface, the process of integration of the complete system has to commence. This must be carried out systematically in a stepwise, calibrated manner to minimize the iterations required for completion of the process and is described in the following sections.

6.4 Software Testing and Integration

Subsequent to finalization of the software design, the software code must be implemented, which may be done manually or using automated tools. As stated earlier, the construction of software is carried out by progressive decomposition of the process level functions into smaller, manageable constituents. Each of these constituents has to be implemented using software such that the complete software architecture is defined as a hierarchy of software modules. Each of the base-level modules can be developed comparatively easily and progressively integrated with other modules in a systematic manner to eventually achieve the desired complete software construction.

Software testing is carried out by executing selected portions of the software using representative input data. The results provide assurance of the developed software conforming to the desired requirements. Despite software design having been carefully carried out and the code being subjected to analysis for compliance to standards, errors do creep into the developed code. Although many potential sources of error may have been detected and eliminated during the analysis process, it is necessary to test the software to verify that it meets the intent of the design requirements. Some characteristics, like timing requirements, can be verified only through testing. Software testing is considered an integral part of the total software development process, which must be done in a systematic manner to confirm that the software code developed carries out the desired functionality at the desired performance levels.

The objective of testing is to ensure all desired functionality has been implemented and there are no errors irrespective of the nature of inputs. The strategy of testing is to devise a series of test cases that would systematically seek out such errors and help correct them. The results expected upon each of the test cases being executed have to be defined; any deviation from expected results implies the presence of errors in the program, and these have to be corrected. The successful completion of the testing process provides assurance that the final accepted code does not contain errors that may surface under some unforeseen situations and cause faulty behavior of the system. However, as stated earlier, software testing can never be exhaustive enough to be carried out in cost-effective time.

The testing process has to be supplemented with the debugging process. If any errors are found during the testing process, debugging has to be carried out to identify the source of error and correct it. The danger of introducing a new error during this process of debugging also has to be guarded against. This may be done by carrying out regression testing of the affected portion of the software each time debugging has been carried out. Regression testing is a repetition of some of the tests carried out prior to effecting a change to ensure that the debug induced change did not have any adverse impact.

The testing of the complete software must be carried out alongside the construction of the software. In general the planning for testing should be carried out before commencement of construction. This includes defining the tests using disciplined techniques as well as identification of the environment required to conduct the tests. Ideally the software should be designed with testability in mind.

Software testing thus begins in the small (base modules) and aggregates toward the large (process of integration). Testing is carried out at "small levels" to confirm each code segment has been correctly implemented. Testing is carried out at "large levels" to confirm the gross software behaves as desired per requirements. The entire software is eventually integrated and tested to the extent practical, in the absence of installation in the final hardware environment.

Thus each of the base-level modules developed has to be individually tested in the defined environment and verified to be providing the desired functionality. The process of testing requires exciting the software unit under tests and monitoring the consequent behavior in terms of outputs generated. This may include use of simulators (with models for closed-loop systems) and software tools to be used by the test team so that all processes and procedures are in place. Simulators should be able to introduce faults and check the response of the software to such faults.

The process of integration of software is carried out at several levels by progressive integration of a large number of tested base-level modules. Testing has to be carried out at defined intermediate stages during such integration of software. The scheme for integration of software should be worked out carefully such that the incremental testing required at each stage is kept to a minimum. Depending upon the architecture of the system it may be possible to carry out testing at intermediate levels for several sets of integrated modules in parallel until the stage is reached when such integrated streams can be brought together.

Testing of software is thus initially carried out at two levels—software unit testing and software integration testing—and these are discussed further in this section. Subsequent to software testing in isolation, the tested software must be installed on the host hardware environment to construct the complete system. During system integration testing, which is the final phase, the complete system is tested including the software tested in conjunction with the hardware, and this is discussed in Section 6.5. The test environment to be used at each of these three levels can be quite different and the specific requirements must be worked out in a planned manner. The techniques used for testing may also be quite different at each of the three levels.

6.4.1 Software Analysis

The use of software in increasingly complex systems led to the challenge of maintaining quality in software. The need to detect and correct flaws in software design at early stages was emphasized earlier.

A static analyzer identifies software code quality issues and detects portions of code that may give rise to unspecified, undefined, or implementation-dependent behavior. It also helps eliminate unnecessary program components that have become redundant (unreachable code) during the program design. It helps with the code review process and will identify segments of code that are nonportable or difficult to maintain. It will identify areas in the code that are too complex and need to be simplified, or may need deeper review.

It enables the process of finding bugs in the software design during the development process without actually executing the software. It is not necessary to wait for the entire development to be over and then subject the software to traditional reliability techniques such as code reviews and functional testing.

A static analyzer assesses source code to identify use of undesirable constructs otherwise legal in the language in use and thereby helps improve the quality of the software development. It automatically checks the source code for compliance with a coding standard, and helps implement a set of rules or best practices set by the organization for example, naming conventions. It can identify configuration and syntax errors in case these have been overlooked by a particular compiler.

A static analyzer determines potential execution paths through the source code, and tracks how the values of variables and other objects could change even across programs. It detects any variable with an undefined value or variables declared but never used. It identifies errors that may arise because of faulty design, including, for example, access beyond an allocated area, use of memory already deallocated, buffer and array underflows, and reading of uninitialized objects. It produces statistics on specific aspects of the source code, for example, unused external definitions or declarations, list of all warnings, and identifiers with similar names.

It generates several metrics that provide a measure of some quantifiable attribute of source code quality. These metrics include *cyclomatic complexity*, depth of nesting, static path count, number of function calls, and number of executable lines, which provide information on complexity, readability, and so on.

The cyclomatic complexity metric (also known as McCabe's number) is a quantitative measure of the number of linearly independent paths in a program code segment (functions, modules, etc.). It is computed by devising the control flow graph of the program. The nodes of the graph represent the primary commands in the program and a directed edge connects two nodes if the command represented by the second node may possibly be executed immediately after the first. In general, the lower this number is for any segment, the better is the quality and the easier it is to understand and test. A high number indicates inadequate modularization or too much logic in one function.

The *cyclomatic number* $V(\mathbf{G})$ of a control flow graph $\mathbf{G}(N,E)$ of a program is calculated as follows:

$$V(\mathbf{G}) = E - N + 2 \tag{6.1}$$

Figure 6.13 shows a simple sample program to demonstrate the concept of cyclomatic complexity.

Figure 6.14 shows the flow graph for this sample program and this has 16 edges and 13 nodes, so the cyclomatic number 5 (see Equation 6.1).

```
#include <stdio.h>
int main(void)
{
        int A=0;
        int B=0;
        int C=0;
        int D=0;
        printf(" Enter A, B, C, D\n");
        scanf("%d %d %d %d ",&A,&B,&C,&D);
        if (A==0)
        {
                printf("First\n");
        }
        else if (B==1)
        {
                printf("Second\n");
        }
        else if (C==2)
        {
                printf("Third\n");
        }
        else if (D==3)
        {
                printf("Fourth\n");
        }
        else
        {
                printf("No match\n");
        }
}
```

FIGURE 6.13
A simple program to illustrate cyclomatic complexity.

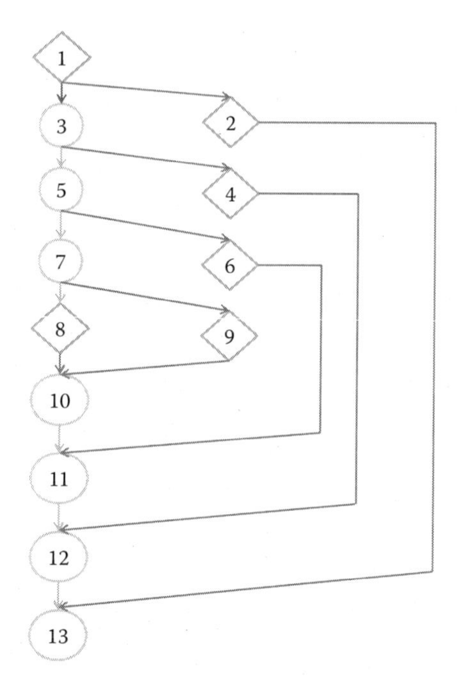

FIGURE 6.14
Directed graph for the program in Figure 6.13.

The more the depth of nesting of IF statements in a program, the more complex and hard it is to understand and it is also more error prone.

For any program, *static path* count is the number of noncyclic execution paths in a program; the higher the number, the more complex the program is to understand and test.

Programs that include a large number of function calls are more difficult to understand and are an indication of a poor design.

Static analyzers [8,9] such as PRQA and LDRA further provide a visualization of the structure of the source code as relationship diagrams to demonstrate function calling, global referencing and include file trees, and control flow structure diagrams for each function.

6.4.2 Software Unit Testing

Unit level testing is generally limited to ensuring that the base-level module (unit), which is implemented as a relatively small piece of software code, exhibits the behavior for which it was designed. To test such behavior it may be necessary to excite the module with the applicable inputs. This may often be done by writing test driver programs to drive such inputs and having additional programs that receive and check the output of the module. Test drivers are software programs written to provide simulated inputs to the programs being tested, to receive the outputs from the same, and provide the test results. Such testing is usually carried out using the development environment used to develop the primary software.

These individual software units have to be tested using low-level tests to verify the small pieces of software code comprising the unit. Each unit must be tested using a predevised corresponding software unit test procedure.

Unit testing is carried out primarily by the use of white box testing techniques to exercise specific paths in a module's control structure to ensure proper flow, correct processing of arithmetic and logic functions, and so forth.

6.4.2.1 White Box Testing

White box testing is carried out based on the knowledge of the internal design details of the software unit. The main objective is to ensure that each microsegment of the software design works properly at the individual level, and working outward steadily, that it also works at the macro level. Such a testing process, which treats the software design as a transparent entity (hence also known as glass box testing), is carried out by devising test cases that exercise each micro segment of the software in a systematic manner.

The test process is carried out by identifying the independent logical paths that the software may take during execution (hence also known as structural testing) and devising tests to check the operation of these paths. The results obtained at the end of the process of execution of a path must match the results that are expected by design. It eventually leads to reasonable assurance that all internal operations conform to the defined requirements. It must be highlighted that in any program of even a moderate length, exercising all paths in an exhaustive manner is not feasible.

It is also required to ensure that each statement, at the source-code level, is exercised at least once, and each branch in the program is also executed at least once.

Further, it is also necessary to check for maintenance of data integrity at each stage, when data is imported into the software unit, during execution of the applicable paths, and when data is exported from the software unit. It is particularly important to check for operation at all applicable boundary conditions.

Some of the kinds of white box tests that are carried out include the following: carrying out tests on all Boolean functions incorporated in each of the program segments by exciting the segment with each possible combination of logic values for each Boolean variable used in the function and confirming that the correct results are obtained.

Each module is represented by a set of flow graphs that depict the pattern of flow of software during the course of execution of the module. A set of independent paths is identified, that is, paths in which some portion of the graph is mutually exclusive. Some such paths may involve flow of software along different tracks depending upon some conditions being met. All such independent paths in a module, including the alternative paths determined by the state of any condition, have to be exercised systematically. (Incidentally this is the same graph and these are the same paths as devised for determining the cyclomatic complexity.)

All loop constructs in the software have to be exercised carefully. Loop constructs may be simple or complex, and tests have to be devised accordingly. Care has to be taken to ensure that loops are exercised properly and tested out particularly at their boundary conditions.

Finally, it is to be ensured that all internal data structures are exercised and all error-handling paths are tested.

6.4.3 Software Integration Testing

Integration testing is the systematic process for constructing a software program structure derived from an assembly of the tested unit software modules and conducting tests to uncover errors as they arise during the assembly process. The process must be carefully planned and documented as the corresponding software integration and test procedure. The outcome of this process will result in assembled and tested software for the computer system. If the system architecture comprises multiple subsystems, the software modules for each subsystem are integrated and tested independently.

The tested unit software modules that are known to be working properly independently may not work as desired when these modules are assembled progressively because of interfacing issues that may arise during this process. The errors may arise because the execution of one module may adversely impact some data that it shares with another module. Data transfer expected from one module to another may not occur due to some mismatch between the designs, particularly when modules are designed by different people. Thus it is required to test explicitly for errors associated with data structures and with flow of data across the interfaces of the modules. Tests must also be designed to search out any errors associated with functionality or with performance requirements.

Most of the testing may be carried out using black box techniques, described in the next section, particularly for functions that require multiple modules for execution. Stub programs may have to be written to simulate portions of software that will be written elsewhere but are not yet ready. Stubs are intended to be only used for testing and are meant to be replaced with linkages to the actual software when it is ready. Smoke testing can be used for a quick check on the important functionality of the software being integrated every time a few modules are introduced into the software build. White box testing can also be used to test data flow paths between modules during integration.

6.4.3.1 Black Box Testing

Black box testing is a method of testing software based on the knowledge of the system requirements at the gross level and with no information on the internal design of the software. This scheme of testing necessitates a study of the requirements and understanding which outputs should be generated in response to the inputs provided under defined execution conditions. The test procedure design entails devising a list of tests such that for a given set of inputs the system should produce the desired outputs.

Unlike as in white box testing, the design of the tests in black box testing is not based on any information about the internal mechanisms of the software programs being tested including its internal code structure, implementation details, and internal flow paths. However, black box testing does not necessarily check all program code flow paths and hence it is treated as an approach complementary to the white box testing process and not as a substitute for it.

Black box testing is used to conduct tests that demonstrate that for all legitimate inputs, the outputs will be such that these match the functional requirements as listed in the defining document. These tests are also used to demonstrate that the system meets all performance requirements listed for it. The technique is also used to find errors in data structures, errors on the edges (viz. interfaces between modules), initialization errors, and so forth.

As emphasized earlier, exhaustive testing of software is not practical. Our attempt in software testing has to be to conduct the least number of tests and yet maintain full coverage as we seek to uncover the maximum number of design errors. We should eliminate redundant test cases similar to others carried out earlier and which do not provide us any additional information about design errors.

Some of the types of black box tests that are carried out include the following:

Equivalence partitioning divides the input domain for a program into valid and invalid classes such that any data from within a particular class will produce the same results. Test cases can be designed to choose one data each from each of such classes and test the program using these. No useful additional information will be derived by carrying out tests on multiple data from any single class.

Boundary value analysis is a strategy to create test cases at the edges of the equivalence classes. Thus test cases are to be created at the boundary value and at just above and just below the boundary value.

Decision table testing (also known as orthogonal array testing) is used when the input domain is small and yet too large to perform exhaustive testing. Representing the correlations between inputs and outputs as a table, the number of tests required can be minimized.

6.4.3.2 Smoke Testing

Smoke testing is a kind of cursory testing of the partially integrated software to confirm that it works properly at an overview level; it does not include detailed and exhaustive testing. It involves integration of minimum usable software and testing of the same to look for any gross-level errors. Smoke testing is carried out, every time some new modules are added to the previous build. The frequent testing at an overview level provides confidence that there are no major errors and it would be possible to proceed with exhaustive testing of the software as and when it is completed. It also implies that errors detected in any iteration are likely to have been introduced in the latest set of software programs integrated into the build. It hastens the building of software for meeting project schedules with short timeframe requirements.

The set of tests intended to carry out smoke testing have to be designed to check for important functionality of the software and look for major errors at an early stage of development such that these do not present hurdles in the subsequent exhaustive testing of the complete software.

6.5 System Integration and Testing

System integration can commence when the complete software has been tested in isolation and, likewise, the complete hardware has also been tested. The complete tested software is installed on the final hardware and testing of the composite system must be carried out. If the system architecture comprises multiple subsystems, the software for each subsystem tested independently is integrated with the corresponding tested hardware and each subsystem is tested separately. On completion of testing of each subsystem, the various computer subsystems have to be progressively integrated and complete system testing carried out.

The process of carrying out the assembly and testing of the various subsystems and, subsequently, to construct and test the complete system is devised as the *system integration and test procedure*. Ideally, the procedure should be defined alongside the process of system decomposition of the system architecture, so as to maintain consistency between the two processes. The procedure should also provide some management information like the size of the test team required and the expected test duration as well as the requirements for physical laboratory infrastructure and resources for carrying out this process.

In the previous sections, strategies and techniques for carrying out testing of software at unit as well as at integrated levels have been discussed. The sections that follow discuss hardware testing and integration, and finally integration and testing of the complete system, including subsystems, if any.

6.5.1 Hardware Testing and Integration

Each individual hardware board has to be fabricated and tested based upon the design information provided. Subsequent to each individual board being fabricated and assembled with all components installed, and after it has passed all quality assurance tests, the board has to be tested as a unit. Thus each hardware unit must be tested by following the corresponding hardware unit test procedure. This may be done manually by exciting each relevant input in a structured manner and checking each output to confirm it behaves as expected. Automated testing can also be carried out using, for example, test equipment built on bed of nails fixtures.

Each board is tested individually until all boards are tested and confirmed to be meeting the design intent. These individual hardware units are assembled together to construct and test the hardware packages as required for each subsystem by following the corresponding hardware integration and test procedure.

Life tests on the hardware can be carried out by accelerated life testing as well as by stress testing. The hardware to be subjected to such tests must be selected carefully to be representative of hardware used in all the subsystems. This process will help evaluate reliability of hardware for each of the subsystems as well as the complete system.

6.5.2 Testing at Subsystem Level

Earlier we discussed in Section 6.3.1 the process of decomposing the system into a hierarchy of subsystems, if necessary, because of the demands of logical structuring of the architecture. The functionality of each subsystem, and of the hardware and software in each subsystem, had to be defined such that all the subsystems put together and functioning in consonance with each other would achieve the total desired functionality specified as requirements to be met by the system.

The process of testing and debugging software modules at the unit level, followed by gradual integration and testing and debugging of the software at the subsystem level was presented in Section 6.4. Likewise, in Section 6.5.1 the process of testing and debugging hardware modules at the unit level, followed by gradual integration and testing and debugging of the hardware at the subsystem level was presented.

Subsequent to completion of the implementation of independent relevant hardware and software packages, these must be assembled in a logical manner to implement each subsystem. Each of these subsystems is implemented by installation of the tested subsystem software on the corresponding tested subsystem hardware. The composite hardware and the composite software (for each subsystem) would have been tested earlier in isolation.

The system integration and test procedure should describe the sequential steps to be followed to integrate the tested subsystem composite hardware and the corresponding tested subsystem composite software. The procedure should also describe the list of functions to be carried out by each subsystem and the specific tests to be carried out to ensure that these functions have been correctly implemented.

Each subsystem should be assembled and tested individually to ensure that the functions assigned to it during the system decomposition process are actually being performed by it.

Although many potential sources of error may have been detected and eliminated during the software analysis and testing process, several characteristics, particularly related to performance, are difficult to verify without actually exercising the software on the target hardware system. This may require establishment of a test environment in which all required inputs can be simulated, including inputs that may originate from some other subsystem. It should also include the means for monitoring all the outputs that may be generated including outputs that may eventually be dispatched to some other subsystem.

Thus, each function to be tested should be described; the inputs to be considered for each test (including boundary conditions) should be identified; and details of the test procedure to be followed, expected results, and actions to be taken if test results do not meet expected values have to be enunciated clearly. The mechanism for simulating all relevant inputs, and monitoring or recording all outputs, and the usage of test tools, if any, also must be identified and arranged for use. This should include simulation of inputs that may actually eventually originate from or terminate in other subsystems in the final composite integrated system.

The testing process has to be completed for each individual subsystem until these are all functioning properly, meeting all functional requirements and performance requirements allocated to each.

6.5.3 System Integration Testing

System integration testing must be carried out when all individual subsystems have been integrated and tested in isolation.

The final stage of this process requires a listing of the sequential steps to be followed to integrate all the subsystems to achieve the complete system. The subsystems should be assembled progressively, in a systematic manner, and tested to eventually construct and test the complete system. The complete system must meet the total set of functional and performance requirements.

Exhaustive test procedures have to be developed to confirm that the integrated system will meet the functional and performance requirements under all operating conditions. The definition of the test plan at this level must be such that each system requirement has at least one test procedure defined to confirm it. More complex requirements may lead to devising multiple test procedures for confirmation of the requirements being met.

Although most testing during integration need be carried out using black box techniques, white box testing can also be used to test communication paths between subsystems.

In addition to testing under normal conditions, it is also advisable to test the system under stressed conditions. Thus tests should be devised that subject the system to unexpectedly large demand on resources, beyond the level expected as per the requirements. The system must still exhibit behavior as required and provide assurance that sufficient design margins are built in to provide desired performance even under unexpected load conditions.

Also, the system should be designed to cope with several kinds of faults. Such faults must be generated and system response reviewed to provide assurance of behavior as designed.

The system must be checked for prolonged periods to demonstrate that it would continue to operate in a stable manner and no errors tend to build up in this period.

The human–computer interface must also be checked during system integration testing. It must be verified that all commands and displays operate exactly as described in the design documents as well as in the user manuals. This includes all types of user interactions: entry of data, execution of control commands, accuracy of data displayed, and display of error messages. It also includes checking of operating procedures, help facilities, use of different operating modes, if applicable, and so on.

6.5.4 System Validation Testing

System validation is the final step in the development life cycle of a computer-based system. This step of checking complete behavior of the system at a black box level is to be carried out by a team independent of the designers. This is also treated as acceptance testing.

The validation process must be carried out in accordance with an exhaustive plan and procedure. At the end of a successful validation in the laboratory, the system is to be installed in the field and wired to the plant signals, sensors, and actuators.

The testing in the development environment (e.g., a laboratory) is carried out using simulators. However, the laboratory environment may be quite different from the actual environment in which the system is to be deployed. Simulators cannot always be designed to replicate the field situation in an exact manner; it may be impractical or expensive to do so. Hence, laboratory testing needs to be supplemented with testing in the actual field environment where the system is finally deployed.

The relevant portion of the validation procedure is required to be repeated under the field conditions and successful completion provides confidence that the system as designed and implemented meets the stated requirements.

NUMERICAL AND ANALYTICAL PROBLEMS

6.1. Elaborate upon the differences between the system integration and test procedure, and system validation procedure.

6.2. Distinguish clearly between verification and validation.

6.3. Write a program on a valid subject option in a course that students can undertake. The subjects are mathematics, physics, chemistry, biology, statistics, and electronics. A student can opt for any four subjects as first, second, third, and fourth subjects. However, some combinations are not allowed, for example, a student taking mathematics as the first subject will not be allowed to take biology, a student taking biology as the second subject will not be allowed to take electronics, and a student taking statistics as the second subject will not be allowed to take chemistry. Determine the cyclomatic complexity of the program.

6.4. Devise a simple program to add two numbers and print "Success" if the sum exceeds 100; determine the difference between the two numbers and print "Yes" if the difference is less than 10. Draw the equivalent flow chart and identify the various paths that must be traversed to achieve complete test coverage.

References

1. ISA S84.01. Functional Safety: Safety Instrumented Systems for the Process Industry Sector. Instrumentation, Systems and Automation (ISA), 2004.
2. IEC 61508 ed. 2. Functional Safety of Electrical/Electronic/Programmable Electronic Safety-Related Systems, International Electrotechnical Commission (IEC), 2010.
3. Benveniste, A., Caspi, P., Edwards, S. A., Halbwachs, N., Guernic, P. L., and Simone, R. D. The synchronous languages twelve years later. *Proceedings of IEEE*, 91(1), 64–83, January 2003.
4. Berry, G. and Gonthier, G. The Esterel synchronous programming language: Design, semantics, implementation. *Science of Computer Programming*, 19(2), 87–152, 1992.
5. Halbwachs, N., Caspi, P., Raymond, P., and Pilaud, D. The synchronous dataflow programming language Lustre. *Proceedings of IEEE*, 79(9), 1306–1320, 1991.
6. MATLAB. Homepage. www.mathworks.com.
7. Esterel Technologies. SCADE Suite. http://www.esterel-technologies.com/products/scade-suite/.
8. Programming Research. Homepage. http://www.programmingresearch.com/.
9. Liverpool Data Research Associates (LDRA). Homepage. http://www.ldra.com/en/.

Index

X